A Trilogy to Define, Plan, an(Contact Center

1) 42 Rules for Using AI in Your Contact Center

Artificial Intelligence (AI) is a transformative force in contact centers, revolutionizing customer interactions. AI tools, including chatbots, virtual assistants, and data analysis capabilities, automate processes, streamline interactions, and offer real-time language translation. These technologies enhance customer experiences by providing personalized services and anticipating needs. The evolving integration of AI holds the potential to create a customer-centric environment, surpassing traditional practices. *42 Rules for Using AI in Your Contact Center* delves into AI's potential impact on customers in the Fourth Industrial Era.

2) 42 Rules for Planning AI in Your Contact Center

Planning for AI in your contact center demands a comprehensive strategy. It goes beyond merely automating agent tasks and requires a focused roadmap to address specific challenges and deliver ROI. Your plan should outline goals, align business and tech objectives, assess your environment, define performance metrics, evaluate AI tech, organize data, and set implementation timelines. *42 Rules for Planning AI in Your Contact Center* provides a roadmap to reduce uncertainties and solve specific problems to yield significant business advantages.

3) 42 Rules to Manage an AI Center of Excellence for Your Contact Center

AI Centers of Excellence (CoE) manage emerging technologies, skills, or disciplines that defy traditional organizational structures. They provide governance, prioritize efforts, and steer businesses away from focusing solely on technology over practical applications. Establishing an AI-CoE for contact centers demands a focused strategy tailored to customer engagement dynamics. It requires meticulous planning, a clear vision, and alignment with broader business goals. Serving as a hub for AI strategy, an AI-CoE ensures initiatives align with business objectives, offers expertise, and guides departments through the complex AI landscape. *42 Rules to Manage an AI Center of Excellence for Your Contact Center* delves into key pillars such as strategic alignment, technology, governance, data, analytics, talent, security, and privacy.

42 Rules to Manage an AI Center of Excellence for Your Contact Center

(Book 3 of 3)

An overview of how to create an artificial intelligence center of excellence focused on your contact center

By Geoffrey A. Best

SUPERSTAR press

E-mail: info@superstarpress.com
20660 Stevens Creek Blvd., Suite 210
Cupertino, CA 95014

Published by SuperStar Press™, a THiNKaha® imprint
20660 Stevens Creek Blvd., Suite 210, Cupertino, CA 95014
https://42rules.com

First Printing: September 2024
Paperback ISBN: 978-1-60773-130-6   1-60773-130-4
eBook ISBN: 978-1-60773-129-0   1-60773-129-0
Place of Publication: Silicon Valley, California, USA
Library of Congress Number: 2024912737

Trademarks

Warning and Disclaimer

Praises for This Book

"In *42 Rules to Manage an AI Center of Excellence for Your Contact Center*, Geoffrey masterfully distills the essence of creating and managing an effective AI CoE. Drawing from my extensive experience in building several Centers of Excellence, which have collectively trained thousands and yielded tens of billions in profit for clients, I can attest to the clarity, conciseness, and precision of this guide. It captures the best practices and well-thought-out processes essential for success in the dynamic world of AI-driven contact centers. This book is an invaluable resource for anyone looking to leverage AI to transform their customer service operations."
Dr. Lucas Root, Ph.D., Community Expert

"I have finally discovered the long-awaited roadmap for mastering the challenges of AI in contact centers. This book helps to uncover the core principles of building an AI Center of Excellence (CoE), empowering leaders and practitioners alike to chart a course toward excellence and innovation."
David Smulders, CEO and Managing Partner, BiTQ

"As an expert in negotiation and critical thinking, I recognize the crucial role of strategic alignment and robust governance in any AI initiative. *42 Rules to Manage an AI Center of Excellence for Your Contact Center* is a masterfully crafted guide that offers comprehensive, practical insights tailored specifically for owners of AI Centers of Excellence. This book will empower you to navigate the complex AI landscape with confidence, ensuring your contact center not only adopts cutting-edge technologies but also achieves sustainable success aligned with your business goals."
Greg Williams, The Master Negotiator & Body Language Expert

"This book provides a comprehensive view of the challenges, opportunities, and approaches needed to address AI in call center applications, which may undergo major disruption due to AI impacting current business models. I recommend that everyone use this as a guide to enable efficiency and understanding in this domain."
Vivek Chhabra, CEO @ 21iQLabs

"An in-depth exploration of the core components of a successful AI CoE. This book equips readers with the knowledge and tools to transform their contact centers into AI-powered hubs for customer satisfaction."
Dr. Mark Hatch, Award-winning CEO and Speaker on Artificial Intelligence

Centers of excellence (CoE) are established to deal with new technologies, skills, or disciplines that don't fit neatly into your organizational structure. An AI-CoE aims to provide governance and prioritize efforts. They help businesses formalize their vision and approach and avoid classic pitfalls, such as a tendency to focus on technology rather than use cases.

Dedication

To Nurcan, my wife, proofreader, and critic; to a technology that continues to evolve without end and challenge all of us; and to my colleagues, family, and friends, who continue to encourage me to document the journey.

Contents

1 Introduction

"Progress cannot be generated when we are satisfied with existing solutions." – Taiichi Ohno, Japanese industrial engineer and businessman who is considered to be the father of the Toyota Production System

Artificial Intelligence (AI) has emerged as a disruptive technological force, especially in the realm of customer engagement and contact centers. The concept of an AI Center of Excellence (AI-CoE) tailored for contact centers represents a strategic commitment to leveraging AI not just as a tool but as a cornerstone of customer experience and operational efficiency. This book aims to be a comprehensive guide for organizations aspiring to develop an AI-CoE specifically for their contact centers. It offers insights into the unique challenges and opportunities in this domain.

Integrating AI into contact centers goes beyond automating responses or streamlining processes. It's about enhancing the customer experience (CX), providing personalized service, and gaining actionable insights from customer interactions. AI technologies such as natural language processing, machine learning, and predictive analytics are reshaping how contact centers operate, promising efficiency and a deeper understanding of customer needs and behaviors.

At its core, your AI-CoE can serve multiple critical functions. First, it serves as the center for AI strategy development, ensuring your AI initiatives align with your business's broader goals and values. It is a hub of expertise and best practices, guiding departments in navigating the complex AI landscape. Secondly, it plays an essential role in talent development, equipping employees with the necessary skills to thrive in an AI-augmented environment. Most importantly, the AI-CoE is a guardian of secure, privacy, and ethical AI use.

Creating an AI-CoE for contact centers involves a focused approach encompassing customer engagement's specific needs and dynamics. It requires a foundational blueprint, meticulous planning, and a clear end-goal vision. This book explores the foundational pillars of such a center, including:

- Strategic Alignment: Ensuring the AI-CoE aligns with the overall business goals and customer service objectives and supports a customer-centric approach in all its initiatives
- Technology and Innovation: Selecting and implementing AI technologies most relevant to contact centers, including chatbots, voice recognition systems, and AI-driven analytics tools
- Governance: Establishing clear governance structures for AI deployment and addressing ethical considerations, such as data privacy and bias in AI algorithms
- Data and Analytics: Centralizing your data science and AI/Machine Learning (ML) resources, promoting knowledge sharing across teams, increasing efficiency in data science initiatives, aligning data science with business strategy, and improving talent acquisition
- Talent and Expertise: Managing the right mix of skills, from leadership, strategic partnerships, AI, and machine learning expertise to customer service acumen
- Security and Privacy: Enhancing capabilities in threat detection, incident response, vulnerability management, risk management, ethics and customer rights, and privacy compliance

Establishing an AI-CoE is not without its challenges. Your business must navigate talent scarcity, technological complexity, integration with existing systems and processes, and security oversight. It is a critical investment in the future of your contact center and a step towards shaping an AI-driven CX.

This book aims to equip leaders, practitioners, and stakeholders with the knowledge and tools to establish and operate an AI-CoE successfully. Focusing on the foundational pillars, this book provides a roadmap for organizations aspiring to lead in the AI era, transforming challenges into opportunities and aspirations into realities.

1 Where to Begin

Establishing an AI-CoE is beneficial and often necessary for contact centers looking to leverage the full potential of AI technologies.

Innovation in business demands an organizational structure that embraces change. It makes the most of the talent and disciplines at its disposal to drive change no matter where it sits in your business. AI is one of the most influential technologies in decades, with the ability to reshape your business. AI can optimize many processes throughout organizations, including your contact center. As AI becomes a permanent aspect of your business landscape, its capabilities must be sustainable to develop and support a culture of innovation and change for new business models and capabilities.

Centers of Excellence (CoE) help create that culture and encourage team members of all disciplines and seniority levels to operate independently from the everyday business of an organization, with all the necessary positions, skills, and autonomy to understand and develop an end-to-end technology or product. This includes roles such as architects, data scientists, people managers, and technologists.

For contact centers integrating AI into their business processes, an Artificial Intelligence Center of Excellence can be that centralized body to strategize, coordinate, and oversee the integration of AI-powered tools and systems into your contact center workflows. The AI-CoE functions as a dedicated hub where organizations can focus on developing, deploying, and optimizing AI solutions tailored to contact center operations' unique requirements. It provides a structured approach to AI adoption, fosters expertise within your business, ensures ongoing technological relevance, manages risks, and fuels innovation. As AI advances, a well-functioning AI-CoE is vital in providing effective contact center operations and serving the evolving demands of customer service.

This is my third book in a series for businesses looking to use, plan, and manage the full potential of AI technologies in contact centers. It explores the process and framework of an AI-CoE for businesses, recognizing that AI has evolved from being important to strategic, shaping your business vision and technology roadmap.

First and foremost, contact centers deal with high volumes of customer interactions on a daily basis. So, optimizing your CX is paramount. Implementing AI in a contact center requires a deep understanding of the technology, its capabilities, and its limitations. The AI-CoE addresses the dynamic nature of AI technology by staying abreast of industry developments and evolving algorithms, evaluating their applicability to the contact center environment, and applying upgrades as needed. This proactive approach confirms that the contact center remains competitive and can adapt to changing customer expectations and industry trends.

In addition to these business advantages, an AI-CoE serves as a focal point for innovation. By encouraging cross-functional collaboration and experimentation, it becomes a breeding ground for novel ideas and AI-driven applications. Contact centers can explore new ways to enhance customer engagement, reduce response times, and optimize resource allocation, which can lead to significant competitive advantages.

Finally, an AI-CoE is instrumental in mitigating risks associated with AI implementation. AI technologies can sometimes yield unexpected results or biases, which may have legal, ethical, or reputational consequences. An AI-CoE establishes strong governance practices, ensuring transparency, fairness, and regulation compliance. It also monitors AI solutions for deviations from expected behavior, alerting the business to immediate corrective actions.

2 What Is an AI-CoE?

An AI-CoE centralizes expertise, knowledge, and resources in a dedicated team with experts in data science, machine learning, and related domains to collaborate on AI projects and initiatives.

An AI Center of Excellence (AI-CoE) is a structured approach that centralizes expertise, knowledge, and resources in a dedicated unit or team within your business. The primary objective of an AI-CoE is to optimize the integration of AI into customer service, including processes, technology, and training. Its goal is to make sure AI in your contact center functions at the highest level of efficiency and effectiveness.

The concentration of AI expertise encourages knowledge sharing, enforces best practices, promotes innovation, coordinates expertise, provisions specialized resources, and facilitates cross-departmental/team collaboration. It brings together experts in data science, machine learning, project management, and related domains to collaborate on AI projects and initiatives. As a result, your AI-CoE can drive innovation, solve complex problems, and develop AI solutions that align with your business's strategic goals.

One of the essential functions of an AI-CoE is knowledge dissemination and skill development. This requires employee training, enabling them to acquire AI-related skills and stay up-to-date with the latest advancements in the field. Having industry knowledge makes certain your AI-CoE team can effectively leverage AI in their respective roles.

Over time, how your AI-CoE operates may transform into a shared services model. As your team discovers best practices, collects feedback, and shares the technology's effectiveness with LOBs, your AI-CoE will be in a solid position to promote enterprise-wide adoption. Your AI-CoE may create new business roles and enable and transform existing roles, such as customer service and sales. Recognize that even in an ideal case, where AI perfectly replicates human behavior, workers will still need to get comfortable collaborating with machines instead of just using them. And in more realistic scenarios, people will also need to learn how to work around AI's unique quirks and limitations.

AI will impact established controls, as well as the process and life cycle of implementing future controls. Your AI-CoE needs to provide effective governance in every phase of the AI life cycle to address its unique challenges and risks, including the following:

- To ensure that AI systems treat your customers fairly and without bias, it is important to have both internal and external checks in place. AI systems that learn from humans and other systems must produce reliable outputs that align with human norms of behavior and are consistent.
- To comply with privacy standards, including your customers' safety, information, choice, and erasure rights.
- To protect against cyber threats.
- To provide clear policies to determine who is responsible or accountable for AI outputs and actions.
- To provide transparency and explainability, explain how data is used and how AI systems make decisions.

3 Why an AI-CoE for Contact Centers?

Contact centers are the lifeblood of organizations. At their core, their primary function is to support customers with services such as sales support, technical support, complaint resolution, or transaction assistance. Properly providing this service will improve customer satisfaction and loyalty. Think about your last interaction with a contact center. We have all experienced calls with excellent support as well as calls that ended in complete frustration.

The evolution of call centers to contact centers has been profound, with shifts from mere call-handling units to sophisticated hubs that manage various customer engagement channels, including web chat, SMS texting, email, fax, voice, and video conferencing. This comprehensive approach lets customers interact with your business on the most convenient platform, enhancing your CX. Contact centers are multifaceted operations extending well beyond traditional customer service. They serve as strategic business units that impact various aspects of a business, from your CX and revenue generation to risk management and data analytics. The diversity of functions underscores the contact center's critical role in your overall business strategy, making it indispensable in a competitive marketplace.

AI brings another level of sophistication with challenges such as limited AI expertise, increasing data complexity, and a need for better tools in AI development. In addition, deploying AI isn't a once-and-done process. It is a series of implementations combining multiple technologies to address a set of use cases spanning functional boundaries. AI, data, and analytics are central to all strategic imperatives. To address these challenges and minimize the chances of an AI project failing, your contact center needs a dedicated business unit to coordinate activities, create standards, and oversee all AI initiatives.

As AI continues to advance, the importance of creating an AI-CoE becomes increasingly evident. For example, in businesses serving multiple lines of business (LOB), each striving to improve efficiency, streamline processes, and leverage technology, there is a high possibility each LOB will work independently. Inevitably, this will unintentionally create redundancies in technology and the use of different vendors, resulting in further discrepancies between departments. To prevent such inconsistencies, it is essential to have an AI-CoE that oversees uniformity across the entire enterprise. Hence, the need for an AI-CoE becomes crucial for all LOBs to be aligned to maximize AI's benefits.

An AI-CoE allows an innovation mindset to form and take hold across your business, a mind shift that won't happen when resources are distributed. Business units typically focus on the near term, and innovation requires a long-term perspective. Innovation also requires a high tolerance for failure that isn't present outside an AI-CoE. As a center for experimentation and innovation, your AI-CoE must encourage cross-functional collaboration and provide a platform for testing new AI technologies and ideas. It is also critical in developing and maintaining best practices and standards. It establishes guidelines for data management, model development, ethical considerations, and governance. This fosters a culture of continuous improvement and allows an organization to identify opportunities for AI-driven enhancements in various aspects of its operations.

If you want to integrate AI into your contact center and align it with the objectives of each LOB, you must ensure your AI-CoE is an integral part of your business model. By enforcing these standards, your AI-CoE can ensure that AI projects are conducted responsibly and comply with regulatory requirements.

4 Define the Value of the AI-CoE

AI-CoEs are focal points for innovation and standardization within an organization. They centralize expertise and resources and promote the development of new AI-driven products and services.

Every contact center aspires to have efficient end-to-end processes and deliver enhanced CX. As such, the value of the AI-CoE should be considered mission-critical to your contact center's technology and business operations. While the technical value focuses on the governance of AI applications and security, the business value of AI may be derived from personalized customer experiences, improved operations, driving cost reductions, and managing risks. AI tools can automate repetitive tasks through chatbots and voice bots, such as ticket classification and fundamental customer interactions. AI can significantly impact your contact center's operational efficiency, innovation capacity, and competitive advantage.

AI technology enables this business value. It offers the ability to analyze vast amounts of data from various sources with predictive modeling. Predictive modeling provides insights for decision-making, forecasting trends, customer behaviors, and potential service issues, facilitating proactive measures and strategic adjustments. Advanced AI applications offer automated and intelligent interactions with your customers without agents.

AI has benefits for both business and technology. The challenge is how to realize its value consistently. An AI-CoE provides mechanisms to manage AI by offering governance. From a technical perspective, establishing an AI-CoE represents a strategic commitment to harnessing and optimizing AI. An AI-CoE fosters innovation, provides oversight of ethical and responsible AI usage, develops talent, and optimizes operations. It is also a custodian of ethical AI practices.

Standardization within contact centers is a focal point for AI-CoEs. AI-CoEs centralize expertise and resources and promote the development of new AI-driven products and services. This centralization optimizes resource allocation and facilitates the sharing of best practices, encouraging a culture of information sharing, continuous improvement, and innovation. AI-CoEs establish norms and practices for developing, deploying, and managing AI technologies and processes within your contact center. They ensure AI technologies are deployed consistently across contact center operations. This uniformity is essential to maintaining the integrity and reliability of AI-driven processes, such as customer interactions through chatbots or automated service recommendations. Consistent deployment helps set predictable outcomes and improves the service quality delivered to your customers.

Establishing standardized protocols and frameworks reduces duplication of efforts in developing and deploying AI systems. Standard protocols offer interoperability between different systems and software used within the contact center, which is essential for integrating data from diverse systems. As AI systems extend to multiple systems in your environment, common mechanisms for accessing and updating data will provide faster integration.

The value of an AI-CoE from standardization, governance, and regulatory compliance will allow AI deployments to establish clear benchmarks and performance indicators, comply with data protection and privacy laws, and facilitate continuous monitoring and quality control of AI applications.

While the value of an AI-CoE is evident, it is important to acknowledge the challenges and considerations involved in its implementation. Establishing an AI-CoE requires significant time, capital, and human resources investment. It requires an obligation to keep abreast of emerging AI trends and technologies to ensure your business remains adaptable and forward-thinking in its approach to AI. It also demands senior leadership's strategic vision and commitment to driving its adoption throughout the organization.

5 View Your CX from the Outside In

An outside-in view of your CX refers to prioritizing and designing your services and products from your customer's perspective.

Your AI-CoE can significantly enhance the perspective of your CX by adopting an "outside-in" approach. An outside-in view of your CX refers to a strategy where your business prioritizes and designs your contact center services, products, and processes by looking at them from your customer's perspective. This approach contrasts with the inside-out view, which focuses on delivering your business's CX vision, strategy, and technology governed by your AI-CoE.

An outside-in perspective is vital for understanding and improving your CX. Adopting an outside-in view in your AI-CoE means leveraging AI technologies for operational efficiency. For example, AI can personalize interactions, predict customer needs, and provide faster, more accurate support, all of which align with the principles of an outside-in approach.

Leveraging AI can be used to personalize interactions, predict customer needs, and provide faster, more accurate support, all aligned with the principles of an outside-in approach. Taking the perspective of outside-in is essential to understanding and improving CX with a focus on:

• Customer-centricity to understand how decisions will impact your customer, rather than solely on internal business objectives or processes
• Mapping cradle-to-grave interactions of your customer journey to identify every touchpoint a customer has with the business
• Providing empathy to understand your

customer's emotions, motivations, and context to help create more meaningful and relevant experiences
- Personalizing customized interactions and experiences based on individual customer data and preferences
- Providing consistency across channels enables your customers to receive a uniform and seamless experience across all communication channels and touchpoints
- Anticipating customer needs and addressing them proactively using predictive analytics and other advanced tools

By promoting this outside-in culture throughout your contact center, your AI-CoE can focus on developing a Customer Technology Platform (CTP), integrating all your customer-facing technology and applications into a single platform to manage and enhance customer interactions and experiences. This synergy between perspective and technology is rooted in a shared goal: to prioritize and improve the customer's needs, expectations, and experiences with a range of functionalities, including:

- Customer Relationship Management (CRM) to track and manage customer interactions, sales, and services
- Contact center technology comprised of telephony systems, automatic call distribution (ACD), interactive voice response (IVR), and other communication technologies that facilitate customer interactions via phone, email, chat, or social media
- Advanced data analytics and reporting to analyze customer data, providing insights into customer behavior, preferences, and trends
- Omnichannel integration of various communication channels like email, chat, social media, and phone ensures consistent CX across all touchpoints
- AI/ML to automate responses, predict customer needs, and personalize interactions. AI in contact centers, for example, can significantly enhance efficiency and customer satisfaction
- Self-service technologies like chatbots, knowledge bases, and self-help portals empower customers to find solutions independently
- Security and compliance with regulatory requirements when dealing with sensitive customer information

6 Centralize or Federate Your AI-CoE

When establishing your AI-CoE, it is important to find a balance between centralization and federation, as both models have their advantages and challenges.

At many companies, LOBs tend to approach planning for AI in a silo, even though there are apparent linkages to upstream and downstream processes and allied functions. Data sourcing and computational methods aren't always defined or consistently used across groups, making data comparisons and cross-functional insights challenging to understand. Access limitations further create barriers to collaboration. Many organizations operate in multi-vendor scenarios with different consultants and systems integrators. Often, the tech stack in data and analytics is vast, with other parts of the same company using different tools. This results in nonstandard outcomes from data and analytic investments. It also impedes knowledge sharing and burdens staff to consolidate the approach from multiple vendors.

Deciding between centralization and federation for your AI-CoE is an important step. Both models offer distinct advantages and challenges. The choice depends on various factors, including your contact center size, culture, objectives, and existing operational models. A centralized AI-CoE operates as an organization's singular, unified entity. It acts as the hub of AI expertise, resources, and decision-making. This is beneficial for tackling complex AI initiatives, enabling consistency in AI strategies and uniformity in methodologies. Centralized models allow for the optimal allocation of resources, including talent and technological infrastructure, in a strong, specialized team with deep AI expertise. It avoids duplication of efforts so that the best expertise is available organization-wide.

On the other hand, centralization can lead to decision-making bottlenecks. A centralized AI-CoE may become a gatekeeper, slowing down AI initiatives in other departments. There is also a risk that a centralized AI-CoE might become disconnected from the unique needs and contexts of different LOBs.

In contrast, a federated model allows LOBs to develop and implement AI solutions tailored to their specific needs and challenges, leading to potentially more innovative and practical applications. When LOBs have more control and ownership over their AI projects, it can lead to increased engagement and motivation. However, a federated model might lead to duplication of efforts and resources as each LOB develops its own AI capabilities.

Meanwhile, as AI gets closer to the core of your contact center, the need for standardization and governance increases. A hybrid approach distributes AI capabilities across LOBs while maintaining a central body that provides guidance and oversight. This allows your AI-CoE to centralize with a federated approach to cross-leverage resources, talent, and technologies.

An AI-CoE functions best when aligned to an organizational matrix. The key here is to strike the right balance between centralization of a function and federated flexibility. Outline which components or activities are most effectively performed by a centralized construct versus those that can be federated in LOBs. The choice between centralizing or federating an AI-CoE is not one-size-fits-all. It requires carefully analyzing your business-specific needs, capabilities, and strategic objectives. Both models have their merits and drawbacks. The best approach may often be a tailored solution that draws elements from both centralized and federated structures.

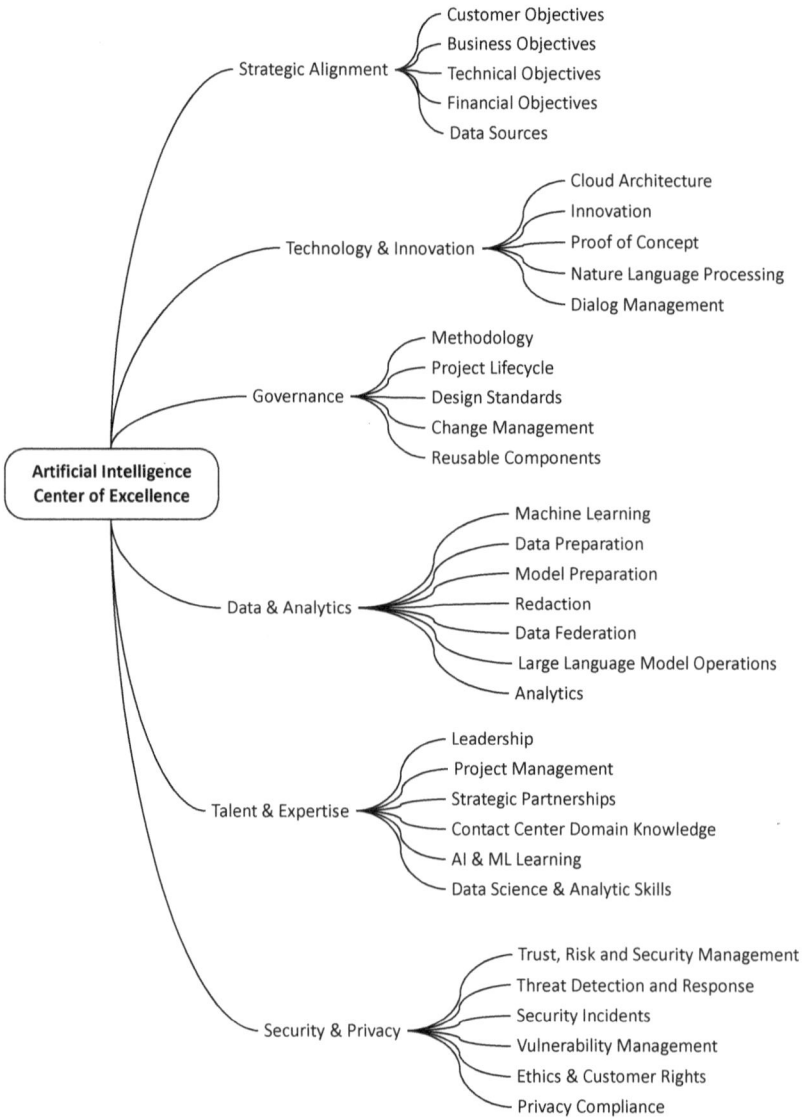

Artificial Intelligence Center of Excellence

- **Strategic Alignment**
  - Customer Objectives
  - Business Objectives
  - Technical Objectives
  - Financial Objectives
  - Data Sources

- **Technology & Innovation**
  - Cloud Architecture
  - Innovation
  - Proof of Concept
  - Nature Language Processing
  - Dialog Management

- **Governance**
  - Methodology
  - Project Lifecycle
  - Design Standards
  - Change Management
  - Reusable Components

- **Data & Analytics**
  - Machine Learning
  - Data Preparation
  - Model Preparation
  - Redaction
  - Data Federation
  - Large Language Model Operations
  - Analytics

- **Talent & Expertise**
  - Leadership
  - Project Management
  - Strategic Partnerships
  - Contact Center Domain Knowledge
  - AI & ML Learning
  - Data Science & Analytic Skills

- **Security & Privacy**
  - Trust, Risk and Security Management
  - Threat Detection and Response
  - Security Incidents
  - Vulnerability Management
  - Ethics & Customer Rights
  - Privacy Compliance

7 Mind Map Your AI-CoE

Mind map your AI-CoE to align with business goals, focusing on enhancing your CX, operational efficiency, innovation, talent, security, and data.

The success of your AI-CoE hinges on cross-functional collaboration and continuous evaluation of its impact on your business through six core components:

• Strategic alignment
• Technology and innovation
• Governance
• Talent and expertise
• Data and analytics
• Security and privacy

Strategic alignment facilitates AI initiatives that are in sync with business goals, focusing on enhancing your CX, operational efficiency, innovation, talent, security, and data. For contact centers, this means aligning AI technologies and projects with key goals such as improving customer satisfaction, optimizing operational efficiency, and driving innovation in customer interactions. Alignment involves a deep understanding of the business's long-term vision and AI's role in achieving it. It requires collaboration across various departments, ensuring that AI solutions are not developed in isolation but are integrated into the broader business strategy.

Technology and integration revolve around deploying advanced AI tools and analytics to optimize contact center operations. Both must continuously adapt and evolve through a structured approach to learning and development. This enables your workforce to leverage AI technologies effectively. Investing in the right technology and infrastructure is critical to supporting AI initiatives. This includes scalable AI and machine learning platforms, cloud computing resources, and tools for developing and deploying AI models.

Governance involves establishing a clear leadership structure and governance model to guide AI initiatives. Effective leadership is essential to aligning AI strategy with overall contact center objectives through clear decision-making, accountability, and resource allocation. This component involves establishing a methodology that defines how AI projects are initiated, evaluated, and managed within the organization. It includes setting up clear policies and guidelines for data usage, model development, and deployment, ensuring that AI applications adhere to data privacy laws, ethical norms, and organizational values. Effective governance also involves setting up oversight mechanisms for AI projects, focusing on accountability, transparency, and risk management. This includes regular reviews and audits of AI initiatives to assess their performance, impact, and alignment with business objectives.

An AI-CoE requires talent and expertise with a team of diverse skills, including data scientists, AI engineers, business analysts, and domain experts. This talent pool is crucial for developing, deploying, and maintaining AI solutions tailored to your contact center's specific needs.

At a time when data breaches and privacy concerns are top-of-mind, security and privacy are indispensable core components. A strong security and privacy framework within the AI-CoE involves implementing advanced cybersecurity measures to protect against unauthorized access, data leaks, and other cyber threats. This includes encryption, access controls, and regular security audits. Equally important is adherence to privacy laws and regulations, such as GDPR, necessitating stringent personal and sensitive data handling incorporating data anonymization and minimal data retention practices.

Finally, data and analytics are a cornerstone of an AI-CoE, providing the foundation for AI-driven decision-making. An effective AI-CoE harnesses this data, transforming it into actionable insights that improve CX and operational efficiency. Establishing robust data collection, storage, and management practices facilitates high-quality, relevant, and up-to-date data availability. Advanced analytics techniques are then applied to analyze customer behavior patterns, sentiments, and preferences, enabling more personalized customer interactions.

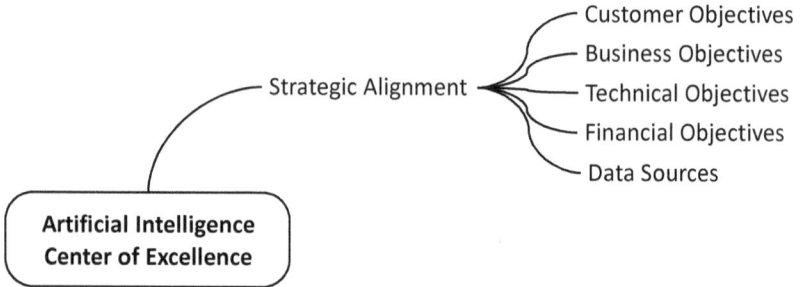

Strategic Alignment
- Customer Objectives
- Business Objectives
- Technical Objectives
- Financial Objectives
- Data Sources

Artificial Intelligence
Center of Excellence

2 Align Your Strategic Objectives

Strategic alignment means associating your contact center's daily activities, projects, and objectives with its mission, vision, and strategy. It bridges the gap between strategy formulation and execution by ensuring that all elements of the organization are working towards a common set of objectives. An AI-CoE for contact centers involves affiliating your objectives, initiatives, and resources with the broader business strategy and goals of:

- Customer Objectives
- Business Objectives
- Technical Objectives
- Financial Objectives
- Sourcing Data

In essence, strategic alignment is about ensuring that your AI initiatives are not just technologically advanced but also directly contribute to enhancing customer service, improving operational efficiency, driving innovation, and ensuring sustainable growth. Strategic alignment with your contact center's operation will enable your AI-CoE's efforts to contribute effectively to your overall mission by aligning customer, business, technical, and financial objectives and data sources through a holistic approach encompassing technology, people, processes, and risk management.

Your AI-CoE is central to aligning contact center objectives and ensuring that AI initiatives are customer-centric and positively contribute to your CX. This alignment involves understanding and integrating customer needs and expectations into developing and applying AI technologies. It involves a deep understanding of your customers' needs, preferences, and pain points by analyzing customer feedback, survey data, and interaction patterns. Your AI-CoE should use this data to identify areas where AI can enhance your CX by providing quick, accurate responses and personalized service, leveraging data analysis to understand and predict customer needs and preferences. It models where AI can improve operations, reduce wait times, and ensure consistent interaction quality, thereby meeting key customer objectives of efficiency, satisfaction, and tailored support.

An AI-CoE views business objectives through strategic contribution and alignment. It models AI innovation to drive improvements in efficiency and scalability by automating routine tasks and handling high volumes of customer interactions, reducing labor costs and response times. It also standardizes advanced analytics and pattern recognition, offering deep insights into the business response to customer behavior and preferences. This aids in tailoring services and improving customer satisfaction, directly contributing to customer retention and loyalty. Furthermore, AI's predictive capabilities assist agent allocation and forecasting, aligning with contact center objectives of maximizing operational effectiveness, enhancing CX, and ultimately driving revenue growth.

Centralizing expertise in your AI-CoE aligns technical objectives, enabling the deployment of advanced, consistent AI solutions across various customer interaction channels. This facilitates continuous improvement and innovation in AI applications, staying abreast of emerging technologies and uniformly integrating them for enhanced service quality and technical robustness.

An AI-CoE helps meet financial objectives by fast-tracking AI implementation, reducing redundant spending, and ensuring efficient allocation of technology investments. Fostering a culture of innovation and continuous improvement drives cost savings through automation and optimized processes, contributing to a higher return on investment.

Standardizing data sources for ingestion into machine learning models by the AI-CoE establishes uniform data collection, processing, and storage protocols, ensuring consistency and quality across the organization. It facilitates the integration of disparate data systems and formats into a cohesive structure, making data more accessible and usable for AI applications. This helps guarantee data fed into machine learning models is reliable, relevant, and reflective of diverse inputs, which is crucial for the accuracy and effectiveness of AI outputs.

8 Align Your Customer Objectives

Alignment with customer objectives is a multifaceted process requiring a comprehensive understanding of customer needs and the technology capabilities to fulfill them.

Aligning your AI-CoE with customer objectives is crucial to promoting cross-functional collaboration and integrating feedback from the service representatives who directly interact with your customers. Coordinating your customer objectives with the broader goals of the contact center is also crucial. This arrangement ensures your AI initiatives contribute to overarching business objectives, such as increasing customer satisfaction and improving service quality. Strategic alignment also ensures compliance with regulatory standards and ethical considerations in AI deployment.

Alignment with customer objectives is a multifaceted process. It requires a comprehensive understanding of your customer's needs and the technology capabilities that can fulfill them. This starts with a deep understanding of what customers expect from contact centers and the expertise to understand industry trends, regulatory changes, innovations, or disruptions. Then, to enhance customer satisfaction, your AI-CoE must regularly gather customer feedback, analyze their interactions, and stay current with their preferences. This alignment can broadly be classified into categories like response time, resolution quality, personalized interaction, and accessibility. It ensures your AI initiatives contribute to overarching business objectives, such as increasing customer satisfaction, reducing operating costs, or improving service quality. Strategic alignment also ensures compliance with regulatory standards and ethical considerations in AI deployment.

A significant aspect of your AI-CoE's role is technological integration, which identifies and integrates the right AI technologies. These could include natural language processing (NLP) for understanding customer queries, machine learning (ML) for predictive analytics, or automated self-service options. Your chosen technologies should align with the nature of customer inquiries and the complexity of issues handled by your contact center.

The human element in AI implementation cannot be overlooked. Training customer service representatives and other relevant staff ensure they can effectively work alongside AI tools. Training should cover the technical aspects of AI solutions and emphasize the importance of human empathy and judgment in customer interactions. The rapid pace of technological advancement in AI can be overwhelming. Regular training programs, attending industry conferences, and partnering with technology experts can help your AI-CoE stay updated and relevant.

Your AI-CoE will need to find the right balance between AI-driven automation and human interaction. While AI can efficiently handle routine inquiries, complex or sensitive issues often require a human touch. Developing a hybrid model that leverages the strengths of both AI and human agents is frequently the most effective approach.

AI implementation can present a significant challenge when it comes to protecting customer data. To maintain customer trust and avoid legal repercussions, your AI-CoE must implement strong data security measures and comply with data protection regulations. It is also critical to consider ethical implications when deploying AI to ensure that AI tools are transparent and fair and do not introduce biases in customer interactions. Regularly reviewing AI decisions for accuracy and fairness and following ethical guidelines will help your AI-CoE maintain the trust your customers expect.

Finally, aligning your customer objectives with an AI-CoE involves staying abreast of emerging AI technologies and trends. This forward-thinking approach ensures that your contact center remains competitive and responsive to changing customer needs.

9 Align Your Business Objectives

Aligning business objects with your AI-CoE will ensure your contact center contributes directly to the broader business strategy.

Aligning the business expectations of your contact center in your AI-CoE is indispensable to achieving operational excellence and long-term success. Contact centers shape your CX, impact brand perception, and ultimately drive business outcomes. Well-managed contact centers must balance efficiency, cost-effectiveness, and customer satisfaction while aligning their objectives with the broader strategic goals of the business.

A primary step in aligning business objectives with your AI-CoE overall goals is to be sure your contact center's activities contribute directly to the broader business strategy, whether it's increasing market share, enhancing customer loyalty, or driving sales. Clear communication of these goals to all levels of your business is crucial for unified direction and purpose.

Operational efficiency is a vital objective for the smooth functioning of a contact center. It involves optimizing workflows, improving processes, and leveraging technology for automation where possible. It also involves efficiently using resources, such as workforce management, over your contact center to help optimally schedule staff, reduce idle time, and manage peak loads effectively. This efficiency not only reduces operational costs but also improves the quality of customer interactions. Your AI-CoE needs to set standards for compliance with legal and regulatory standards, a critical aspect of contact center operations. This includes data protection laws, consumer rights regulations, and industry-specific guidelines.

Another objective is effectively managing contact center costs while maximizing the return on investment (ROI). Your AI-CoE needs to oversee technology investments, workforce management, and process improvements, balancing cost-cutting measures with the ability to invest in quality service and technology.

High-quality customer service is a non-negotiable aspect of aligning business objectives. Quality CX requires quick response times, effective resolution of inquiries or issues, and high customer satisfaction. Your AI-CoE should recommend quality assurance measures, like call monitoring and feedback mechanisms, in addition to recommending regular training and development programs for agents. To align your business objectives, your AI-CoE should also focus on hard skills like product knowledge and soft skills like empathy and communication.

Integrating advanced technologies like AI, ML, and data analytics can significantly enhance the contact center's capabilities. Your AI-CoE ensures standards for these technologies effectively automate routine tasks, assist agents with real-time insights, and facilitate personalized customer interactions. AI-driven tools can analyze vast amounts of data to identify trends, predict customer behavior, and optimize resource allocation. Coordinating the technological integration across the enterprise improves efficiency and provides a competitive edge.

AI data analytics is also crucial in meeting business objectives, agent performance, and operational efficiency. By analyzing this data, your AI-CoE can identify areas for improvement, track the effectiveness of strategies implemented, and make informed decisions. Continuous improvement based on data-driven insights ensures that your business objectives evolve in line with changing customer expectations.

The contact center industry is subject to rapid change due to techno-logical advancements and shifting consumer preferences. Your AI-CoE should foster innovation and allow contact centers to adapt quickly to these changes, embrace new technologies, and experiment with novel service delivery models. This adaptability is crucial for maintaining align-ment with the business's increasing demands.

10 Align Your Technology Objectives

Your AI-CoE must be involved in building and integrating the infrastructure and technology backbone for the modern contact center.

The alignment of your contact center's technical objectives involves a comprehensive approach encompassing robust infrastructure, advanced technology integration, data security, scalability, high-quality communication services, and continuous innovation. This alignment addresses your contact center's need to maximize the usage of on-premises or in-cloud computing and highly skilled personnel with the best technical solutions to your customers' challenging, complex, and evolving business needs. By focusing your AI-CoE on aligning technical objectives, it can leverage AI technology for reuse across different business segments with a repository of solutions and reusable components, sharing best practices and knowledge base.

The integration of infrastructure and technology is the backbone of the modern contact center. Your AI-CoE should not only be involved in building this backbone but also own its management and ongoing progression. It should provide the oversight to scale up or down based on demand with the flexibility that ensures your contact center can handle peak periods efficiently without compromising service quality. This includes the integration of telephony infrastructure with software such as Customer Relationship Management (CRM) to access customer data during calls.

Your AI-CoE may recommend cloud-based solutions for their scalability and flexibility, allowing contact centers to adjust resources quickly per fluctuating customer interaction volumes. Moreover, your AI-CoE must establish advanced analytics and reporting capabilities for network and server performance. Proper oversight will eliminate dissimilar tools, inefficient performance testing, and incorrect analysis.

Primary tools are real-time analytics for performance management and detailed reporting that offer insight into contact center key performance indicators (KPI) for measuring end-user experience. This will also assist your AI-CoE in understanding customer interactions, agent performance, and operational efficiency. KPIs will also help make data-driven decisions to enhance service quality and operational efficiency.

Staying abreast of emerging technologies and trends for technical advancement is essential. This might include exploring customer-impacting technologies like 5G, the Internet of Things (IoT), or advanced speech analytics. Being open to innovation and continually updating your technological stack will keep your contact center ahead in a competitive market.

AI and automation are transforming contact center operations. AI is utilized for predictive analytics, natural language processing, and machine learning to enhance customer interactions. For example, chatbots and virtual assistants can handle routine queries, while AI-driven analytics can provide agents with real-time insights and recommendations during customer interactions.

In addition, the increasing trend of video calls in customer interactions requires an infrastructure to ensure high-quality voice and video services. Your AI-CoE should recommend investing in high-capacity solutions with a robust network infrastructure to handle high bandwidth requirements for clear, uninterrupted voice and video communication.

Finally, contact centers must have adequate disaster recovery and business continuity plans. Your AI-CoE must plan to deal with unexpected events like system failures, natural disasters, or other disruptions. The plan should include data backups, alternative communication channels, and contingency plans to ensure operations can continue with minimal downtime.

11 Align Your Financial Objectives

Aligning financial objectives requires a strategic approach to contribute to the broader financial success of your business.

As you align your customer, business, and technical objectives for AI in your contact center, you will need to align clear financial strategies with your business's broader objectives. Aligning financial objectives with your AI initiatives requires a strategic approach encompassing precise goal setting, effective budget management, ROI measurement, integration with organizational finance, stakeholder engagement, continuous improvement, technology utilization, risk management, and benchmarking. By focusing on this approach, your AI-CoE can not only meet its financial objectives but also significantly contribute to your business's broader financial success. This involves setting specific targets and benchmarks that need to be achieved to align with the AI-CoE strategic objectives. Start by analyzing your contact center's historical financial data to identify trends, patterns, and areas that need improvement. This will provide insights into your business's past performance and help you set your financial objectives.

Next, consider external factors that may impact your contact center's financial performance. This could include market conditions, industry trends, regulatory changes, and competitive landscape. Understanding these external factors will help set realistic and achievable financial goals. These goals should be specific, measurable, attainable, relevant, and time-bound. Align the goals with your business's overall financial objectives, including costs for resources, technology, staff training, and other expenses related to the AI-CoE's operations.

Determine a budget to achieve your strategy, including specific quantitative targets that must be committed to align with the customer, business, and technical objectives for integrating AI in your contact center. In addition to quantitative targets, consider qualitative goals aligned with your strategic objectives. These could include customer satisfaction, brand reputation, and other non-financial goals that contribute to the integration of AI.

Foster stakeholder engagement by regularly communicating financials to key stakeholders and demonstrating how they contribute to your business's financial objectives. This is crucial for ongoing support and funding. Integrate your AI-CoE's financial planning with your business's overall financial strategy, including synchronization with your fiscal calendar, budgeting processes, and financial reporting systems.

Be prepared to adapt strategies in response to changes in your business environment, market trends, or your AI-CoE's operational needs. This will enable you to identify any deviations and take corrective actions if needed. Monitoring progress will help ensure that the financial goals remain aligned with the business objectives and that the organization stays on track toward achieving its targets.

Establish realistic timelines for achieving the financial goals of your AI-CoE. Consider the resources available, market conditions, and the complexity of the goals when setting timelines. Setting realistic timelines will help ensure your financial goals are achievable and effectively aligned with the customer, business, and technical objectives. Regularly assess the financial risks associated with your AI-CoE's operations, including market, operational, and compliance risks.

Stay informed about industry best practices or other organizational CoEs. Apply these learnings to improve financial performance by comparing your AI-CoE's financial performance with industry benchmarks to gauge competitiveness and efficiency.

12 Identify Your Data Sources

Identify a landscape with a data ownership framework, access control, and quality standards to source data for AI and ML.

One of the fundamental aspects of your AI-CoE is identifying the data landscape of your contact center. To start, thoroughly audit your data from the originating system, including data formats and quality. Sourcing data for AI and ML is not solely an IT endeavor. Your AI-CoE should foster collaboration with contact center managers, business analysts, data scientists, and other relevant stakeholders to identify your data source. The inventory should identify available data types, such as customer interaction summations, historical records, and real-time metrics. This includes descriptions of data fields, their meanings, and the context in which they are used. Well-documented data makes it easier for data scientists and analysts to work with the data effectively.

With a clear data ecosystem overview, your AI-CoE should establish a robust data governance framework. This framework should define data ownership, access controls, and quality standards. Depending on the industry, governance should ensure data is accurate, consistent, and compliant with relevant regulations, such as GDPR or HIPAA.

A crucial aspect of sourcing data for AI and ML is integrating data from diverse sources. Your AI-CoE should devise an integration strategy that allows seamless data flow between systems. This may involve using Extract, Transform, and Load

(ETL) processes or modern data integration platforms. Data security and privacy are also paramount when sourcing data for AI and ML. Your AI-CoE should work closely with your legal and compliance teams to establish protocols for securing sensitive customer data. This includes encryption, access controls, and audit trails to track data usage.

In many cases, raw data from contact center interactions may require enrichment and preprocessing before it can be used effectively for AI and ML models. Your AI-CoE should define best practices for this stage, which may involve text analytics, sentiment analysis, or data augmentation techniques to enhance the data quality. Maintaining data quality is an ongoing process. Your AI-CoE should implement continuous monitoring and quality assurance mechanisms to identify and rectify issues promptly. Data quality dashboards and alerts can be instrumental in this regard.

Addressing ethical concerns and bias in AI and ML models is an increasingly critical aspect of data sourcing. Your AI-CoE should implement guidelines and methodologies to identify and mitigate bias in data, ensuring fairness and equity in AI-driven decisions. As your AI and ML capabilities evolve, it should plan for scalability and future-proofing data-sourcing practices. This may involve exploring cloud-based data storage and processing solutions that can handle growing data volumes and new data types.

Finally, your AI-CoE should invest in training and skill development for its team members and other relevant personnel. Data scientists, engineers, and analysts must be well-versed in best practices for sourcing data and effectively working with AI and ML models. This includes understanding the data landscape, establishing assertive aegis, supporting data security and privacy, and fostering collaboration across departments. With a well-defined approach to data sourcing, your contact center can unlock the full potential of AI and ML, driving improved CXs and operational efficiency.

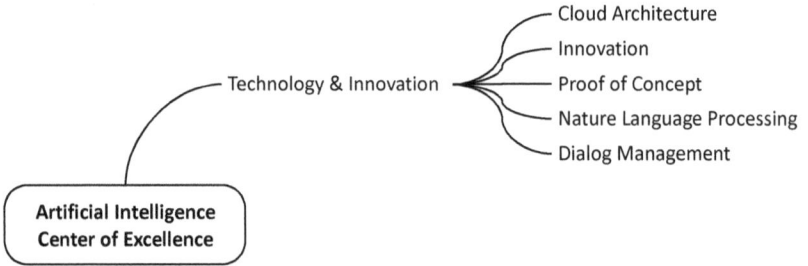

Technology & Innovation
- Cloud Architecture
- Innovation
- Proof of Concept
- Nature Language Processing
- Dialog Management

Artificial Intelligence
Center of Excellence

3 Technology and Innovation

"It's easy to come up with new ideas. The hard part is letting go of what worked for you two years ago but will soon be out of date." – Roger von Oech

Continuous research and evaluation of emerging technologies are fundamental for identifying innovation opportunities in your contact center. This includes creating technology strategies that align with your business's goals and objectives, as well as identifying and addressing potential risks and challenges associated with technology adoption. As a centralized hub specializing in fostering expertise and promoting best practices, your AI-CoE needs to lead this effort and be proficient in various technology domains, including cloud architecture, innovation, proof of concept, NLP, and dialog management.

In cloud architecture, your AI-CoE must guide the design of scalable, secure, and efficient solutions, ensuring your cloud strategy aligns with your contact center's overall goals. It establishes standards and frameworks for cloud adoption, guides the selection of cloud service providers, and oversees the implementation of cloud-based solutions. By focusing on emerging cloud trends and technologies, your AI-CoE ensures that your contact center stays at the forefront of cloud innovation, leveraging the cloud's full potential to drive business growth and efficiency.

Innovation is at the heart of your AI-CoE's mission and, as such, needs to operate as an incubator for new ideas and technologies, creating an environment that encourages experimentation and creative thinking. This involves identifying emerging technologies and market trends, evaluating their relevance, and considering their potential impact on your contact center. It must foster partnerships with external entities, such as startups, academic institutions, and research organizations, to bring fresh perspectives and expertise into the innovation process.

Your AI-CoE should validate new technologies or approaches by demonstrating their feasibility and potential impact before full-scale implementation. This involves setting up proof of concept (PoC) programs, testing, and evaluating the results to ensure the new technology meets the desired objectives under real-world conditions. Managing PoC projects helps mitigate risks associated with adopting new technologies.

In NLP and dialog management, your AI-CoE is strategic in developing and refining technologies that enable machines to understand, interpret, and respond to human language effectively. This role must be centrally positioned if your business seeks to enhance customer interactions through intelligent and conversational AI. Your AI-CoE must guide the development and implementation of NLP technologies that enable machines to understand, interpret, and generate human language. It must engage not only the technical aspects of NLP, such as machine learning algorithms and language models, but also the integration of NLP into customer service applications to ensure NLP solutions are developed ethically, respect privacy, and avoid biases.

Dialog management is a subset of NLP and is particularly relevant in contact centers and customer service. Your AI-CoE must oversee the development of dialog systems with the ability to conduct natural and effective conversations with users. It involves a design review of conversational flows, response strategies, backend databases, and integration of contact center applications. Your AI-CoE must certify these dialog systems are continuously improved based on user feedback and evolving business needs to make sure your contact center AI is user-friendly, context-aware, and capable of handling a wide range of customer inquiries and tasks.

13 Architect Your Cloud

At a strategic level, your AI-CoE must carefully assess the requisite cloud infrastructure necessary to support your potential AI use cases.

Setting standards, developing best practices, and publishing guidelines are essential for your AI initiatives to align with your contact center's broader technology strategy. Your AI-CoE should not have day-to-day operational responsibilities, nor should it be a project management organization. It must act in a consultative role to establish an overarching cloud architecture. It must oversee your business's cloud adoption for AI to be sure the right skills and structure are in place for a successful cloud deployment. At a strategic level, your AI-CoE must carefully assess the requisite cloud infrastructure to support your AI use cases. This means selecting a cloud platform with the right computational power, scalability, and data storage requirements to meet your AI needs. By choosing the most suitable cloud environment, your AI-CoE lays the foundational framework for deploying your AI solutions.

After selecting a cloud platform, your AI-CoE will need to provide oversight to monitor cloud-based AI systems for compliance with these standards, addressing risks related to data breaches, ethical AI practices, and misuse of AI technologies. This includes establishing and enforcing policies and standards that confirm your cloud-based AI applications adhere to regulatory requirements, data privacy norms, and security protocols. It also consists of the ongoing responsibility of performance monitoring and cloud optimization to assess the AI systems for their computational efficiency, resource utilization, and cost-effectiveness. Your AI-CoE must continuously evaluate the performance of these systems, identifying areas for improvement and making necessary adjustments to ensure that the AI solutions are not only performing as intended but are also cost-efficient and scalable. This governance role is critical in maintaining the trust and integrity of your AI systems, particularly in sectors where sensitive data in the cloud is handled.

Vendor management is another area where your AI-CoE should participate. Many AI solutions involve third-party vendors dependent on cloud service providers that may be outside of your cloud ecosystem. Your AI-CoE must provide guidelines for these vendor relationships, ensuring that third-party services and products align with your business's AI objectives and cloud architecture requirements. This consists of overseeing contract negotiation, managing service level agreements (SLAs), and ensuring that vendors deliver value that meets your expectations.

Finally, your AI-CoE should be responsible for driving innovation and future-proofing your contact center's AI cloud initiatives. This means staying abreast of emerging trends and advancements in AI and cloud computing, evaluating how these can be integrated into the existing architecture, and planning for future expansions or upgrades. It warrants AI solutions are effectively integrated into the cloud environment to harness the power of AI, drive innovation, and maintain a competitive edge. By keeping an eye on the horizon, your AI-CoE validates that your contact center not only leverages current technologies to their fullest but is also prepared to adopt new and potentially disruptive technologies as they emerge.

14 Foster Innovation

Fostering innovation means encouraging experimentation with new AI technologies and methodologies in pilot projects and proof-of-concept initiatives.

One of the primary objectives of your AI-CoE is fostering a culture of innovation and harnessing AI technologies. As a centralized body governing the evolution of AI initiatives, your AI-CoE must be responsible for defining where AI can add value. For instance, your AI-CoE might focus on implementing an AI-powered interactive virtual assistant (IVA) leveraging NLP and ML to interact with customers conversationally, provide immediate responses to queries, and offer personalized assistance.

Fostering innovation means encouraging experimentation with new AI technologies and methodologies. Innovation may involve pilot projects or proof-of-concept initiatives that explore new AI applications in the contact center domain— for example, AI-driven predictive analytics for customer behavior. As a result, your AI-CoE becomes a core of technical know-how that can be shared with different LOBs, facilitating cross-functional teams to work on AI projects. Such collaboration between AI experts and customer service professionals can lead to more effective and user-friendly AI solutions.

Innovation in AI requires managing risk, including ethical concerns, data privacy issues, and potential biases in AI models. Your AI-CoE needs to identify these risks and develop strategies to mitigate them. This is particularly important in contact centers where AI systems interact directly with customers, making it crucial to maintain transparency, fairness, and privacy.

To determine if innovation is successful, your AI-CoE needs to demonstrate its value. This means setting performance indicators and metrics to track the effectiveness of AI projects. In the context of a contact center, these metrics could include customer satisfaction scores, resolution times, or cost savings. By quantifying the impact of AI, your AI-CoE can justify continued investment in your AI initiatives and guide future innovation efforts.

Forming a clear strategic direction means more than just technological advancements to drive your AI projects. They must contribute to your contact center's long-term objectives. Part of setting a strategic direction is establishing best practices and standards for consistency and reliability across different LOBs and following a framework for data governance, model development, and ethical AI usage. This is important to maintain the quality and integrity of AI systems, particularly in sensitive areas such as customer data handling in contact centers.

An AI-CoE often leads the effort to upskill an existing workforce and attract new talent with AI expertise. This effort may include talent development and skill enhancement through training programs, workshops, and collaboration sessions to build AI literacy across your organization. For instance, your AI-CoE may train customer service representatives on AI tools or educate the IT team on the latest AI technologies.

Promoting innovation demands a culture that encourages new technologies and tolerates failure. Innovation thrives in environments where teams from different groups are empowered to explore novel applications of AI, challenge existing paradigms, and pursue creative solutions without concern for a lack of success. Steering the strategic direction of AI initiatives while fostering innovation requires a culture of experimentation, continuous learning, strategic partnerships, and strong oversight. Collectively, these traits drive value and position your contact center for long-term success.

15 Proof of Concept

Your AI-CoE is responsible for the entire PoC process, from ideation to the pilot phase, including the technical aspects, architecture, algorithms, timelines, milestones, and resource allocation.

An AI proof of concept (PoC) is a small-scale project designed to demonstrate an AI solution's feasibility and practical potential in solving a specific problem or improving a process. It is a preliminary test to evaluate an AI implementation's viability, technical aspects, and real-world applicability before committing to full-scale development. Identifying a problem or a process to be improved is usually a collaborative effort between your AI-CoE, your contact center, and your business. Each PoC should be consistent with your contact center objectives as well as feasible from a technical perspective.

Once a problem is identified, your AI-CoE should undertake extensive data collection and analysis. This involves gathering relevant data critical for training and testing the AI models. Data needs to be cleaned, preprocessed and made ready for use in AI algorithms. Your data's quality, quantity, and relevance directly influence the PoC's success. Concurrently, your AI-CoE must work on selecting the appropriate AI methodologies and technologies based on the problem's specific characteristics, the data's nature, and the desired outcomes. Methodologies might range from machine learning and natural language processing to more complex neural networks or deep learning techniques, depending on the complexity and requirements of the project.

The development of an AI data model is an iterative process. The prototype is tested and retested against a subset of data. This iterative process is basic to understanding the model's accuracy, adjusting parameters, and areas for improvement. The iterative nature of model development allows your team to fine-tune the model and retrain it with additional data if necessary. It is also during this phase that your AI-CoE must confront and address potential challenges such as overfitting, underfitting, biases, or favoritism in the model.

Next, your AI-CoE must establish that your proof of concept adheres to ethical standards and complies with relevant regulations and policies. This involves assessing the AI's decisions for fairness, transparency, and accountability, especially in sectors where AI decisions might have significant consequences. Your AI-CoE must address privacy concerns, ensuring data handling and processing comply with data protection laws and organizational policies.

If, and when, your AI model demonstrates promising results and meets the predefined objectives in a controlled environment, your AI-CoE must prepare for a pilot phase. This involves deploying your AI solution in a real-world setting, albeit in a limited scope. The pilot phase is critical for understanding how your AI solution performs in actual operational conditions and provides insights into any practical challenges that might arise. It also allows end-users to interact with the AI solution, providing valuable feedback that can be used to refine the model further.

To complete your PoC, your AI-CoE must document the process from ideation to the pilot phase. Documentation should contain the technical aspects of the AI solution, such as model architecture and algorithms, as well as project management aspects like timelines, milestones, and resource allocation. Comprehensive documentation is an absolute necessity for scaling the AI solution beyond the PoC stage, ensuring that the knowledge and insights gained are preserved and can be leveraged for future initiatives.

16 Natural Language Processing

Create a blueprint for NLP applications, identifying where NLP can be effectively implemented by considering current needs and future scalability.

Natural Language Processing (NLP) is a field of AI focused on enabling computers to understand, interpret, and generate human language in a way that is both meaningful and useful. By leveraging computational linguistics and machine learning techniques, NLP algorithms can analyze text and speech data to extract insights, automate tasks, and facilitate communication between humans and machines.

Oversight of NLP is complex. From strategic planning to continuous improvement, your AI-CoE must be sure NLP technologies are effectively planned, implemented, managed, and leveraged to achieve your contact center goals. It must supervise NLP applications' quality, compliance, and efficacy, particularly in domains such as search engines, chatbots, voice assistants, social media analysis, machine translation, sentiment analysis, text summarization, and more.

To manage this undertaking, your AI-CoE must develop a blueprint for NLP applications, identifying where it can be implemented while considering current needs and future scalability. This strategic perspective enables NLP solutions to align with your broader contact center goals and integrate seamlessly with existing IT infrastructure and business processes. It enables NLP solutions to align with broader contact center goals and integrate with existing IT infrastructure and business processes. This viewpoint involves establishing and enforcing standards and best practices for NLP applications, including data handling, model training, and algorithm development guidelines.

To advance your NLP applications, your AI-CoE will need to innovate continuously. It must lead initiatives to explore new NLP techniques and tools to enhance your contact center's capabilities. This means keeping abreast of the latest developments in the industry and actively participating in original research. By doing so, your AI-CoE will contribute to the growth of organizational knowledge and keep your business at the forefront of NLP technology.

NLP applications often involve processing large volumes of personal and sensitive data. Your AI-CoE must safeguard that applications comply with legal and ethical standards, including data privacy laws and AI ethics guidelines. One of the main challenges for your AI-CoE is the quality and diversity of data. Data is the fuel for NLP models but is often noisy, incomplete, inconsistent, or biased. Your AI-CoE must warrant that the data used for training and testing NLP models and systems are representative, relevant, and reliable.

The performance of NLP applications must also be continuously monitored. Your AI-CoE must set KPIs and benchmarks to evaluate the effectiveness of NLP solutions and identify issues early to ensure NLP systems are delivering the desired outcomes. Quality assurance is a continuous process that involves regular testing and fine-tuning of NLP models to maintain high accuracy and reliability.

Effective governance of NLP requires regular engagement with stakeholders, including management, end-users, and external partners. Your AI-CoE has to act as a bridge between these stakeholders, ensuring clear communication and alignment of expectations. This involves presenting the value and potential of NLP applications to your management, gathering feedback from end-users to improve user experience, and collaborating with external partners for technology sharing and integration.

Often, the implementation of NLP solutions involves collaboration with external vendors and technology providers. Your AI-CoE must participate in selecting vendors, managing contracts, and supervising the integration of external technologies with internal systems. This includes evaluating the technical capabilities of vendors, ensuring compliance with organizational standards, and managing ongoing relationships to leverage external expertise effectively.

Dialog Management

Dialog management in AI involves orchestrating conversations between humans and machines by determining when to listen, when to respond, and how to maintain context, in a seamless and meaningful interaction. It encompasses techniques such as intent recognition, context tracking, and response generation to enable natural and effective communication between users and AI systems.

Governance of dialog management is a linchpin in your AI-CoE strategy. As such, it must provide a structured approach by orchestrating the myriad aspects of customer interaction, ensuring that every communication aligns with the highest standards of efficiency, consistency, and satisfaction. Through your AI-CoE, your business can harness the full potential of its contact centers, leveraging advanced technologies while maintaining a focus on human-centric service. The many-sided role of your AI-CoE in dialog management encompasses establishing best practices, technology integration, training and development, continuous improvement, compliance, and cross-functional collaboration.

Establishing best practices and standards for dialog management means defining communication protocols, response time benchmarks, and quality assurance measures. By setting standards and best practices, your AI-CoE can provide consistency in customer interaction, which is a cornerstone for maintaining high customer satisfaction and loyalty. Setting standards involves guiding the selection, integration, and optimization of these tools in dialog management. These tools include various AI solutions such as NLP, sentiment analysis, and automated response generation. Your guidance must ensure the chosen technology aligns with your organization's goals and customer needs. Using AI tools ethically is crucial for providing seamless and personalized communication with customers across different channels.

As AI/ML technologies continue to advance, your AI-CoE must align these technologies with your organization's strategic objectives and customer service ethos. Beyond selection, it must maintain these tools' ethical and effective implementation, balancing the drive for efficiency with the imperative to maintain personal and empathetic customer interactions.

Monitoring and continuous improvement are also central to the dialog management framework. By implementing strong monitoring mechanisms, your AI-CoE can track the effectiveness of dialog management strategies, as well as analyze customer feedback, call resolution times, and the performance of AI-driven interactions. This ongoing analysis is not merely about oversight; it is a springboard for continuous refinement, enabling the identification of areas for enhancement and innovative dialog management practices in alignment with customer needs and expectations.

Compliance and ethical considerations are also increasingly at the forefront of dialog management, particularly with the growing reliance on AI and data analytics. Your AI-CoE must establish stringent guidelines for data privacy, security, and the ethical use of AI, ensuring that dialog management practices comply with regulatory requirements and uphold the highest ethical standards. This governance aspect is critical in maintaining customer trust, particularly in an era where data breaches and ethical lapses can severely impact brand reputation.

Lastly, your AI-CoE must facilitate cross-functional collaboration, recognizing that effective dialog management is not the purview of the contact center alone but a cross-departmental endeavor. By fostering collaboration between IT, customer service, marketing, and other relevant departments, your AI-CoE will set the foundation for dialog management strategies to be integrated with the organization's broader objectives. This collaborative approach will deliver a cohesive dialog management that meets the current demands of customer service and is poised to adapt to future challenges and opportunities.

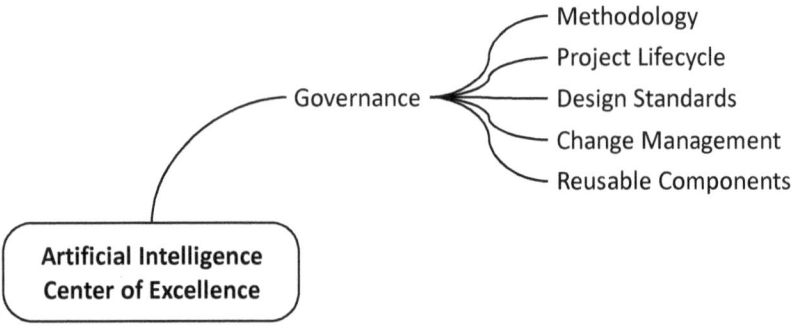

Methodology
Project Lifecycle
Governance — Design Standards
Change Management
Reusable Components

Artificial Intelligence
Center of Excellence

4 Governance

> "I'm increasingly inclined to think that there should be some regulatory oversight, maybe at the national and international level, just to make sure that we don't do something very foolish." – Elon Musk

Effective governance in an AI-CoE for your contact center is about more than just enforcing rules and policies. It is about creating a dynamic ecosystem where strategies, technology, people, and processes coalesce to deliver exceptional customer service. As AI and other technologies continue to reshape the landscape of customer interactions, the governance structures of your AI-CoE must evolve to stay ahead of these changes, ensuring that contact centers are efficient, compliant, innovative, and customer-centric. It must manage diverse and sometimes conflicting stakeholder expectations and navigate complex regulatory landscapes. However, these divergences also present innovation opportunities, improved CX, and operational efficiency.

Overall, AI-CoE governance refers to the control of strategic direction, business cases, investment, and risk management through a formal governing body dedicated to successfully implementing AI in your contact center. It provides expertise in managing governance practices and supporting projects. One of your biggest challenges in establishing a governance team will be determining the governance structure or the balance between centralized and federated responsibilities. A fully centralized governance model is where decisions, discussions, and development are funneled through one group of people; on the other hand, a fully federated model is one where individual business teams have complete freedom to do whatever they want without internal approvals or coordination. Most companies land somewhere in the middle, with corporate culture dictating most of the decision process. Effective governance involves reaching across departmental boundaries to various stakeholders, including employees, management, customers, and vendors, to ensure your AI-CoE's strategies and operations are transparent and accountable.

Your AI-CoE has the opportunity to encourage a culture that embraces AI and its impact on customers and employees. Leadership plays a crucial role in driving this cultural shift, making sure that staff are well-trained in customer service skills, handling new technologies, and understanding policies.

Developing comprehensive policies that govern the use of technology, data handling, and customer interactions is essential. Policies must comply with legal and ethical standards, such as GDPR for data protection and other industry-specific regulations. Governance frameworks help establish mechanisms for continuous monitoring of service quality and performance metrics. This allows your contact center to operate efficiently and meet predefined service level agreements (SLAs).

Your AI-CoE is also responsible for identifying and managing risks, especially in areas like data security, operational disruptions, and technology implementation through regular audits and updates to risk management strategies. As contact centers increasingly integrate AI's capabilities, governance becomes even more pivotal in steering them toward operational excellence, compliance with regulations, and achieving business goals. This includes the ethical and responsible use of AI principles around transparency, fairness, non-discrimination, privacy, and accountability in AI applications. AI systems often handle sensitive customer data, prioritizing their security and privacy, and your AI-CoE needs to be sure your contact center complies with data protection laws and implements robust data security measures.

18 Create a Methodology

Create a rigorous methodology as a strategy that guides how AI is conceptualized, executed, and managed.

Creating a methodology for AI in contact centers involves a cyclic process of defining goals, collecting and preprocessing data, training models, evaluating performance, and iterating the process based on insights gained. Forming this methodology can be complex. It requires guiding the implementation and management of AI projects in a structured framework. This framework offers a standardized way of working that enables consistency in handling AI projects across your business. It helps organize and streamline the processes involved in AI development, from initial concept to deployment and maintenance. It brings order and clarity to what can otherwise be a complex and challenging endeavor.

A methodology helps create a common language and understanding among team members and stakeholders. In an AI-CoE, methodology is not just a set of tools and processes. It is a strategic asset that guides how AI initiatives are conceptualized, executed, and managed. Adopting a rigorous methodology enhances the competence of the AI-CoE team and builds credibility both within the organization and in the external market. It demonstrates a commitment to excellence and reliability in AI project execution. It helps ensure that AI projects are aligned with organizational goals, are executed efficiently, and deliver tangible business value.

Your methodology should incorporate industry best practices into the AI project management process. This includes risk management, quality assurance, and ethical and legal standards compliance. A methodology enhances process efficiency by providing a clear framework and set of procedures. It helps streamline project activities, from ideation to deployment, ensuring that resources are optimally utilized and timelines are adhered to. In AI projects, which can be complex and multifaceted, a methodology helps in breaking down the project into manageable parts. This includes clear guidelines for each project phase, such as requirement gathering, design, development, testing, and deployment.

AI projects have unique risks, including data privacy concerns, algorithmic bias and favoritism, and technical challenges. A robust methodology aids in systematically identifying, assessing, and mitigating these risks. Your methodology should be adaptable to accommodate the rapidly evolving nature of AI technologies and market demands. It should allow for iterative development and continuous learning, which are crucial in AI projects. A suitable methodology will include tools and metrics for measuring the performance of AI projects. This enables continuous improvement and helps demonstrate the value and ROI of AI initiatives to stakeholders.

Governance of methodologies often incorporates mechanisms for documentation and knowledge sharing. This is essential in AI projects for continuous training, refining algorithms, and improving practices over time. Your chosen methodology should align with the organization's culture and values. For instance, an Agile methodology might be more appropriate if an organization values agility and innovation. A Waterfall methodology provides stage gates for organizations that want a more structured approach.

Methodologies often include monitoring, evaluation, and feedback mechanisms, which are essential for continuous improvement in AI projects. Learning from past projects and refining approaches based on experience and feedback is crucial in the fast-evolving field of AI. As your organization's AI initiatives expand, a robust methodology can provide the scalability to manage increasingly complex projects. It allows for the growth and evolution of the AI-CoE without compromising the quality and coherence of your AI initiatives.

19 Project Lifecycle

Your AI-CoE must govern your AI project lifecycle to identify how AI is deployed in your contact center environment.

Methodology is about "how" a project is managed, including approach and best practices, while a project lifecycle is about "what" stages the project goes through, including the sequence of steps. Methodologies can be applied across various projects, not just in AI, and often reflect your organization's broader project management culture. On the other hand, project lifecycles are tailored to the specific phases of a particular project, including unique aspects of AI development in this context. Methodologies can be more rigid or flexible depending on the chosen framework. For example, an Agile methodology is more adaptive than Waterfall, whereas the project lifecycle is generally linear and sequential, especially in complex AI contact center projects.

Governance of your AI project lifecycle helps to identify the details of how AI is deployed in your contact center environments. This includes gathering oversight about your business goals and what resources are needed at each step. Ultimately, this translates into maintaining control of the project more effectively. This represents a comprehensive and systematic process through which AI projects are conceptualized, developed, and maintained.

A project lifecycle typically begins with the Initiation Stage, where opportunities for AI application are identified, and the project's feasibility is assessed, involving stakeholders to define its scope and objectives. This stage is followed by the Planning Stage, where re-sources are allocated, risks are assessed, and a detailed project plan is developed. The Design and Development Stage should focus on collecting and preparing data, developing and training AI models, and creating prototypes. This phase is crucial as it lays the foundation for the AI solution.

Once an AI model is developed, your project moves to the Testing and Validation Stage, where the model's performance is rigorously evaluated against set metrics. Its alignment with project requirements and stakeholder expectations is verified. This stage often involves iterative improvements based on feedback and test results. The Deployment Stage follows, involving the integration of the AI solution into existing systems, careful planning for the rollout, and transitioning the solution to a live environment.

The project lifecycle then progresses to the Monitoring and Maintenance Stage, which is vital for the project's ongoing success. Here, the performance of the AI solution is continuously monitored, and regular updates and maintenance are carried out to ensure its effectiveness and relevance. Feedback loops are established for continuous improvement.

Finally, the lifecycle concludes with the Project Closure Stage, where the project's overall success is evaluated, lessons learned are documented, and stakeholder approval is obtained. In the Post-Project Stage, knowledge and insights gained from the project are transferred across the organization, and possibilities for scaling or adapting the solution for other uses are explored. Governance of the project lifecycle is not just a linear progression but allows for feedback and iterative improvements, reflecting the dynamic and evolving nature of AI technology and its applications.

20 Standardize Your Design

AI introduces ethical challenges like bias, privacy, and decision transparency, requiring design standards to ensure your AI is effective, reliable, and ethical.

Design standards serve as a comprehensive set of guidelines and best practices to ensure consistency, quality, and efficiency in developing, deploying, and maintaining AI solutions across your organization. By encompassing technical, ethical, and operational considerations aligned with your organization's broader goals, design standards will enable your AI to be effective, reliable, and ethical. To begin, your AI-CoE needs a deep understanding of AI technologies, including machine learning, natural language processing, and neural networks. AI heavily relies on data. Understanding data collection, storage, processing, and usage is necessary to ensure data quality, integrity, and security. Finally, you must address how AI solutions integrate with existing technology infrastructure and workflows in the contact center environment. Your design standards need to prioritize identifying and mitigating biases in data sets to ensure your data models are ethical in a manner that supports the specific needs and challenges of contact center operations. The result is AI solutions that are fair, unbiased, and technically sound but also practical, user-friendly, and comply with legal regulations. Design standards should also provide for your user experience, ensuring that AI implementations are user-friendly and accessible to a diverse range of users, including those with disabilities.

Design standards in your AI-CoE must provide a structured framework with best practices emphasizing the importance of high-quality, accurate data for model training, testing, and validation protocols to ensure accuracy, reliability, and robustness. A structured framework often encourages collaboration between disciplines, such as data science, domain expertise, and ethics. This interdisciplinary approach can lead to more holistic and practical model training.

Part of your AI-CoE's role is establishing clear metrics for evaluating AI performance and continuous monitoring protocols to ensure the AI systems function as intended. This will make your AI solutions scalable and maintainable over time, with clear updates, upgrades, and potential decommissioning guidelines.

While design standards support consistency, allowing room for innovation and adaptation is still necessary. Design standards must be flexible enough to accommodate creative solutions and evolving technologies. AI remains an evolving field, and design standards are not static; they must evolve with changing technologies, market trends, and consumer preferences. Include principles for continuous learning in your design standards, where models are regularly updated and improved based on new data and feedback. This will enable your AI-CoE to keep pace with technological advancements and emerging best practices, which in turn will enable your designs to stay relevant and effective.

Lastly, your AI-CoE's design standards must support your company's mission and help achieve its objectives. Establishing a feedback loop where designers and other stakeholders can provide input on the standards helps in continuous improvement. By involving stakeholders in the standard-setting process, you ensure buy-in and usability.

21 | Manage Change

Change often requires significant modification in processes, roles, and technologies. Your AI-CoE must emphasize the importance of change and support the adoption of AI.

Change management is the systematic approach to facilitate and manage the process of implementing significant changes within your business. It requires setting standards and best practices to plan, implement, manage, and sustain change effectively. Your AI-CoE's role is to emphasize the importance of change management and support your employees' adoption of AI. Change often requires significant modification in processes, roles, and technologies, and you will need a comprehensive plan to manage the steps of bringing AI to your business. This includes allocating resources, developing timelines, and setting critical milestones.

Start by developing a change management strategy with clear objectives and specific, measurable goals for your AI initiatives. These goals may be improving efficiency, enhancing CX, or generating new insights. Whatever your goals, your AI-CoE should promote change management to obtain buy-in for changes, addressing risks and vulnerabilities associated with your future state, and ensuring stakeholders are prepared for a new status quo. Change management should include defining the scope of transformation, building a business case, and gaining sponsorship from the right leaders. It also includes identifying pain points where change has already occurred or targeting where it is needed.

As a hub of expertise and best practices, your AI-CoE needs to guide the strategic integration of

AI across various departments to align with organizational goals and mitigate the risks associated with technological adoption. It orchestrates the change process by setting standards and policies for AI use, ensuring consistency and compliance across the organization. Your AI-CoE should also facilitate training and development programs, equipping employees with the necessary skills to adapt to AI-driven workflows and technologies.

Assess how new AI technologies will transform existing business processes and identify any benefits from the change. Determine who will be affected by the AI implementation and understand their concerns, needs, and attitudes towards AI. Establish a clear and consistent communication plan to inform your stakeholders about the AI initiatives, their benefits, and the expected changes.

Lastly, your AI-CoE must act as a bridge between technical AI teams and other business units, fostering cross-functional collaboration and communication to ensure that AI initiatives are well-understood and supported throughout the organization. By closely monitoring AI projects, your AI-CoE can gain valuable feedback and insights, allowing for continuous improvement and optimization of AI strategies. It can play a central role in managing the cultural shift towards a more data-driven and AI-integrated approach, addressing concerns and resistance that may arise during the transition. Overall, your AI-CoE ensures the adoption of AI is a structured, strategic, and smooth process. This involves emphasizing the importance of change and enabling people to adapt to AI by using technology in new ways, often requiring significant changes in processes, roles, and technologies.

22 Reuse AI Components

The role of your AI-CoE in component reuse is to encourage the appropriate and best use of resources.

Reuse of AI components refers to the practice of leveraging existing algorithms, models, libraries, or frameworks to expedite the development of new AI systems or enhance existing ones. By reusing proven components, developers can save time and resources, improve system reliability, and focus on higher-level tasks such as customizing algorithms for specific applications or fine-tuning model parameters. Reuse may also leverage pre-built AI solutions and components to minimize coding and redundant system development. This approach is strategic for maximizing efficiency, ensuring consistency, and fostering innovation.

Your AI-CoE's role in reuse is to encourage the appropriate and best use of resources by centralizing support for shared reusable assets. For instance, in a model-based architecture, AI functional units can be pre-trained to be reused directly or extended by incorporating new conditions and corresponding actions.

Promoting the reuse of components can accelerate the development of AI applications in your contact centers. For example, reusing standardized AI models pre-trained or previously developed for specific tasks, such as NLP or speech recognition, ensures your contact centers don't have to start from scratch for every new implementation. To facilitate reuse, emphasize creating modular AI components that can be easily integrated into different contact center applications. This approach allows for the easy addition, removal, or updating of components without disrupting the entire system. It also fosters standardization by promoting the adoption of established best practices, frameworks, and tools across projects. By leveraging reusable technology components such as pre-trained models, libraries, and pipelines, your AI-CoE can ensure consistency in development approaches, data handling methods, and deployment strategies. This standardization streamlines processes, reduces duplication of effort, and facilitates knowledge transfer among team members, ultimately enhancing efficiency and promoting a unified approach to AI development for your contact center.

Your AI-CoE should establish a central repository for AI components where reusable artifacts such as algorithms, models, datasets, and code snippets are stored, organized, and made accessible to developers across the organization. This repository provides a single source of truth for AI assets, enabling easy discovery and retrieval of relevant components for new projects. By promoting collaboration, sharing, and version control, a repository facilitates reuse by empowering teams to leverage existing solutions, promote consistency, and reduce redundant efforts.

The reuse of AI components can significantly improve operations by enhancing various aspects of customer service and support. For instance, AI Models like OpenAI's GPT (Generative Pre-trained Transformer) series or Google's BERT (Bidirectional Encoder Representations from Transformers) can be reused across various NLP tasks such as text classification, sentiment analysis, or language translation. Frameworks like Rasa or Microsoft Bot Framework provide reusable components for building conversational AI agents, including dialog management, intent recognition, and entity extraction. These components enable developers to create sophisticated chatbots more efficiently. Most importantly, feedback and learning from previous implementations is part of the continuous improvement process promoted by your AI-CoE.

Artificial Intelligence
Center of Excellence

Data & Analytics

- Machine Learning
- Data Preparation
- Model Preparation
- Redaction
- Data Federation
- Large Language Model Operations
- Analytics

5 Data and Analytics

"You can have data without information, but you cannot have information without data."
– Daniel Keys Moran

The objective in managing data and analytics should be to compress your ingestion timeline and reduce maturity costs. To achieve this objective, your AI-CoE must be a centralized resource for data science and AI/ML, promoting knowledge sharing across teams, increasing efficiency in data science initiatives, aligning data science with business strategy, and improving talent acquisition. It must also promote data and model literacy across the business, standardize data science workflows and tools, and support and prioritize innovation initiatives.

Specific to your contact center, your AI-CoE needs to ensure data is ready for machine learning in Large Language Models (LLM), data federation is effective, and relevant metrics are defined.

Machine learning algorithms use data to optimize customer interactions, predict trends, and automate responses in conversational voice and chatbots. Algorithms can be employed for sentiment analysis, customer segmentation, and predictive analytics. Your AI-CoE must ensure the continuous improvement of these algorithms through iterative training and refinement. This involves collecting diverse data sets, implementing feedback loops, and staying updated with the latest advancements in machine learning techniques.

Data preparation refers to data quality and readiness for effective use in AI applications. This includes aspects such as data cleanliness, labeling, structuring, and integration. Ensuring data readiness is crucial for the accuracy and efficiency of AI models. Your AI-CoE must establish stringent data governance policies, perform regular data audits, and maintain data lineage documentation to ensure high-quality data is available for analysis and machine learning purposes.

Metrics are essential for evaluating the performance of AI initiatives. Key performance indicators should include customer satisfaction scores (CSAT), resolution times, and accuracy of AI responses. Your AI-CoE needs to establish a framework for continuously monitoring these metrics. This involves tracking performance over time and interpreting these metrics in the context of business goals and your CX.

LLMs are machine learning models that aim to predict and generate plausible language. They are used for deep learning and are instrumental in understanding and generating human-like text, making them invaluable in handling customer queries and providing information.

Managing LLMs involves regular updates and training with relevant data to ensure the models are aligned with the specific needs of the contact center. Ethical considerations and biases in the model must also be monitored and addressed in a framework that interprets the metrics in the context of your contact center goals and your CX.

Data federation refers to integrating data from disparate sources, presenting it as a single, unified data source. This might involve combining customer data from various platforms and databases in contact centers. Your AI-CoE should facilitate data federation to enable a holistic view of customer interactions across different channels. This involves using advanced data integration tools and ensuring compliance with data privacy regulations. Each federated component is interdependent, and your AI-CoE must ensure they align with your contact center's overarching strategic goals for your business.

23 Manage Machine Learning

Your AI-CoE must continually focus on emerging ML technologies, so your conversational agents can personalize your CX.

Data in machine learning is a foundational element for training algorithms, providing the raw material from which these algorithms learn patterns, make predictions, and derive insights. Data must be meticulously collected and labeled for supervised learning tasks. This enables the machine to adjust its internal parameters and improve its performance over time, adapting to new data inputs and evolving in accuracy and efficiency. Ensuring your machine learning algorithms are implemented effectively to optimize customer interactions, predict trends, and automate responses is a prime objective of your AI-CoE. This comprehensive stewardship involves strategy, governance, and continuous improvement. Stewardship ensures these advanced technologies are operational and aligned with your business's customer service objectives. It also advances ML applications that automate tasks, improve accuracy, and provide insights into customer behavior. While ML applications may provide customers quick service, they may also have limitations. Many chatbots in use today rely on rule-based systems or traditional ML algorithms or models. Generative AI (GAI) has the potential to advance customer service significantly by leveraging LLMs and deep learning techniques to understand complex inquiries and offer to generate more natural conversational responses. GAI can consider the history of your customer's interaction from data in multiple platforms to provide them with information delivered in their preferred tone and format.

Your AI-CoE must continually focus on emerging ML technologies, so your conversational agents can personalize your CX. By setting the standards for training on vast datasets of customer interactions, advanced ML algorithms can identify common issues and queries, enabling them to provide quick and accurate responses. This ensures that conversational agents are trained to handle a variety of customer temperaments and interaction styles to foster a personal and empathetic connection.

ML applications require systematic governance to perform optimally. Your AI-CoE must provide this governance to establish guidelines for data usage, algorithm training, and performance benchmarks. By overseeing the data fed into these algorithms, your AI-CoE ensures they learn from accurate, relevant, and diverse data sources to avoid biases and inaccuracies in customer interactions. Performing regular audits and adjusting these algorithms are part of your AI-CoE's role to maximize the integrity and efficiency of AI systems.

Predictive trend analysis of historical data to forecast future customer behavior and trends enables your contact center to address potential issues proactively before they arise. Your AI-CoE must set up the algorithms and continuously refine them to improve their predictive accuracy. This can lead to better-targeted marketing campaigns and a more streamlined customer service experience.

An essential aspect of your AI-CoE's mandate is to facilitate the continuous learning of machine learning models. This involves initial training and ongoing updates to the models as new data comes in. Your AI-CoE must oversee this iterative process, ensuring that conversational agents are updated with the latest product information, customer service practices, and regulatory changes, thereby maintaining their relevance and effectiveness.

Ethical considerations and regulatory compliance predominate when applying machine learning in customer interactions. Your AI-CoE must enforce ethical guidelines and compliance measures to ensure customer data is handled responsibly, privacy is maintained, and all interactions comply with industry standards and regulations.

Your AI-CoE will face several challenges in overseeing ML applications, including ensuring data privacy, avoiding algorithmic bias, and maintaining a balance between automated and human interactions. You must address these challenges through robust policies, transparent practices, and continuous governance.

24 Prepare Your Data

Data readiness involves ensuring that data is available, accurate, clean, and in a format suitable for your AI models.

Data is the underpinning upon which AI algorithms learn, adapt, and make informed decisions. Sourcing data for AI involves meticulous collection, curation, and preparation of information from various sources, such as databases, sensors, social media platforms, and more. Your AI-CoE has a multifaceted role in data preparation, from advocating for data quality to fostering a data-driven culture and managing data infrastructure. Its oversight is vital for any business seeking to leverage AI effectively. Ensuring data readiness enables the development of robust AI solutions and lays the foundation for sustainable AI-driven growth and innovation.

Data readiness involves making sure data is available, accurate, clean, and in a format suitable for your AI models. This encompasses aspects such as data collection, data quality, data governance, and data infrastructure. AI systems can only function optimally with proper data readiness, leading to unreliable outputs and diminished value from AI investments.

Your AI-CoE needs to be a champion for high data quality standards. It must establish guidelines for data acquisition, processing, and management, ensuring that data used for AI models is reliable and valid. This includes implementing rigorous data cleaning and preprocessing procedures, establishing protocols for data labeling, and advocating for consistent data entry practices across departments.

In brief, your AI-CoE must govern data to warrant its responsible and ethical use, including who has access to it, how it can be used, and how to handle sensitive information. This oversight is instrumental in establishing a data governance framework that addresses data privacy, security, and compliance.

Your AI-CoE must foster a data-driven culture within your business, educating and training employees on data quality and data-driven decision-making. It involves conducting workshops, seminars, and training sessions to enhance the data literacy of the workforce and ensure that all employees understand the value of data and its role in AI applications. Data often resides in silos within organizations, making it challenging to harness for AI. AI-CoE facilitates data accessibility and integration by implementing systems that allow for the seamless flow of data across different departments and platforms. This integration is vital for developing comprehensive AI models that require diverse data sets from various sources.

To effectively implement AI initiatives, your AI-CoE must oversee the development and maintenance of a robust data infrastructure. This includes investing in the right hardware and software tools, such as cloud services, data warehouses, and data lakes, ensuring the organization has the necessary technology stack to process and store large volumes of data efficiently. Analytics can help contact centers evaluate behavioral patterns, improve customer satisfaction, and enable more efficient staffing and resource allocation.

In summary, your AI-CoE is responsible for guiding the business in developing strategic methods for collecting data that align with its AI objectives. This involves identifying relevant data sources, both internal and external, and using effective data collection techniques. The goal is to ensure that the data collected is representative, unbiased, and substantial enough to train robust AI models. To achieve this, the AI-CoE needs to identify customer KPIs, build data models, and publish predictive analytics dashboards. By doing so, the company can make more informed decisions based on data-driven insights.

25 Model Preparation

Model preparation requires meticulous data development and management. When correctly performed, it can significantly enhance the efficiency and efficacy of contact center operations. Preparation must consider multiple factors, including accuracy, interpretability, training time, and computational efficiency. Algorithm selection is contingent upon the problem type (classification, regression, clustering, etc.) and the nature of the data. Common algorithms include decision trees, neural networks, support vector machines, and ensemble methods like random forests and gradient boosting machines.

Your AI-CoE must navigate the intricacies of algorithm selection, data management, model training, and deployment to prepare your model for descriptive, predictive, diagnostic, and prescriptive analytics. Model preparation involves choosing appropriate algorithms, training models on relevant and diverse datasets, and regularly updating models to reflect changing patterns in customer behavior. It must occur in a structured and strategic process to ensure that AI models align with your business objectives and conform to ethical standards.

Feature engineering involves creating model variables that serve as the foundation upon which algorithms learn and make inferences. It is the method by which raw data is transformed into a format better suited for machine learning models, effectively enhancing the performance of these models in making predictions or decisions. Your

AI-CoE must identify which features most identify customer behavior and likely outcomes. This might involve text analysis of customer service transcripts, social media sentiment analysis, or transaction data trend analysis. Once features have been engineered, they become the variables or inputs for models. These variables are what the model will use to learn from the training data. The quality of these features directly impacts the model's ability to learn and make accurate predictions or prescriptions.

Your AI-CoE should collaborate with IT teams to integrate models into existing systems. This involves technical considerations like API development, data pipelines, and computing resources. The goal is a seamless integration where models can provide real-time predictions and prescriptions within the contact center's operational framework. It must also engage with stakeholders, including management and contact center agents, to ensure that models meet practical needs and that there is organizational buy-in for AI initiatives.

Before deployment, models must be validated and rigorously tested to generalize new data. Your AI-CoE must oversee the establishment of metrics such as accuracy and precision. Post-deployment, your AI-CoE must continuously monitor performance. Models may drift over time as customer behavior changes, so your AI-CoE must establish protocols for regular updates. When updates are applied, it may result in retraining models with new data or tweaking algorithms to adapt to new patterns. Your AI-CoE needs to supervise retraining to ensure models comply with ethical standards and regulatory requirements. This involves transparency in how data is used, allowing customers to opt-out, and ensuring that models do not discriminate against any customer segment.

Lastly, your AI-CoE is responsible for training contact center staff to interpret and act upon model outputs. It must provide ongoing support to foster a culture of innovation and encourage ongoing research and development. This could involve exploring new data sources, experimenting with cutting-edge algorithms, and staying abreast of AI and machine learning advances.

26 Redact Your Data

Consider the critical aspects of redaction, such as what sensitive data needs to be de-identified, which technique to apply, and whether the redacted data maintains its value for analysis downstream.

Your contact center's confidential information must be protected through data redaction using a strategic framework with defined responsibilities, standards, risk mitigation, and regulatory compliance. Data redaction is a technique that masks sensitive data by removing or substituting parts of it to protect personally identifying information. Redaction can be full, partial, or lookup-based, depending on the sensitivity of the data and the need to retain its utility for downstream analysis. Redaction techniques include complete and partial redaction with customizable options like redacting specific parts of data and masking characters to replace sensitive information.

Consider the aspects of redaction, such as what sensitive data needs to be de-identified, which technique to apply, and whether the redacted data maintains its value for analysis downstream. Redaction is irreversible; therefore, careful consideration is needed before applying it to ensure the redacted information is of value. Partial redaction can provide information without revealing the complete details, maintaining a balance between privacy protection and data utility.

As such, your AI-CoE must be responsible for formulating and disseminating policies that delineate the parameters for data redaction. Implementing data redaction policies across the entire data lifecycle is paramount, and policies must consider legal requirements, industry standards, and organizational objectives. These policies will help limit unauthorized access by restricting unnecessary information and reducing the risk of leaks or data breaches. Policies must also govern applying data redaction at different stages, such as when acquiring data, after completing projects, or before archiving.

Central to the worth of your AI-CoE is harnessing the technological solutions that facilitate data redaction. Tools for advanced data masking, encryption algorithms, and access control mechanisms are key in guarding against unauthorized access or disclosure of confidential data. Your AI-CoE must continuously evaluate emerging technologies and best practices, ensuring that your contact center remains at the forefront of data protection measures. It must also monitor legislative developments, regulatory guidance, and emerging threats by proactively updating policies and practices to align with your contact center's evolving requirements. Moreover, your AI-CoE must use benchmarking technology and knowledge sharing to provide insights and best practices.

Ultimately, your AI-CoE is not just about managing existing data but also about exploring new ways to protect data for AI, including experimenting with advanced data analytics techniques, researching new data sources, and innovating in data visualization. It is the custodian of trust for safeguarding sensitive data in your contact center. It provides strategic oversight, operational guidance, and a continuous improvement ethos for data redaction governance.

27 Federate Your Data

To federate your data, your AI-CoE must act as a strategic coordinator, governance body, technology advisor, educator, and innovator.

Data federation refers to integrating and managing data from diverse, autonomous, and heterogeneous sources to provide a unified, consolidated view of the data without physically centralizing it. Your AI-CoE must set clear objectives for data federation that align with the contact center's overall goals. It must create a strategic roadmap for how data federation should be implemented, considering the specific needs and challenges of the contact center environment.

Your AI-CoE needs to act as a strategic coordinator, governance body, technology advisor, educator, and innovator. Its role is crucial in ensuring that data federation efforts are aligned with your business goals, technically sound, compliant with regulations, and effective in enhancing your contact center's performance and customer service capabilities. It must bring data together from data stores using different storage structures, access languages, and APIs.

Once data is federated, contact center applications can access all the data sources on demand and on the fly, not in batch. Moreover, data is only accessed and integrated when an application queries it. Thus, the data is not stored in an integrated manner. It remains in its original location and format, able to operate independently and be used outside the scope of data federation.

Regardless of how and where data is stored, it should be presented as one integrated data set through a federation layer that acts as an intermediary and translates queries from the AI to each data source. Your AI-CoE must define the Unified Query Interface to provide a single point of access for querying the data, often using standard query languages like SQL.

This implies data federation involves transformation, cleansing, and possibly even enrichment of data. In addition, the autonomy of the data stores allows them to operate independently in that they can be used outside the scope of your contact center applications. Your AI-CoE should advise selecting appropriate technologies for data federation, including data integration tools, databases, and AI applications that suit the specific needs of the contact center. It must play a role in designing the infrastructure to support data federation, ensuring scalability, reliability, and efficiency.

For example, imagine an AI system designed for customer service in a contact center. This AI needs to access federated data to retrieve customer information from a CRM database, a knowledge base for troubleshooting steps, and real-time chat logs from various communication channels. In this scenario:

- The AI queries the federated system through a unified interface.
- The federation layer translates this query to fetch relevant data from the CRM, knowledge base, and chat logs.
- Results from these sources are aggregated to provide a response to the customer.

Your AI-CoE must provide oversight for performance optimization, including data latency, consistency, and scalability. Specifically, your AI-CoE should recommend the approach for caching frequently accessed data and distributing query loads across multiple data sources to prevent any single source from becoming a bottleneck.

28 Large Language Models Operations

The LLM framework encompasses ethical guidelines, compliance with regulatory standards, and policies that ensure its responsible use.

Large language models (LLM) are sophisticated AI systems and a driving force behind a significant shift in AI and ML. By reshaping NLP and pushing the boundaries of language understanding and generation, LLMs have become a crucial part of our technological landscape. Built on advanced deep learning architectures, such as transformers, and trained on vast amounts of text data, LLMs have mastered the nuances of human language. As LLMs become more prevalent in your contact center applications, the need for efficient and effective operational practices increases.

LLM Operations (LLMOps) is a set of practices and techniques that provide a strategic approach to managing and deploying LLMs at scale. Your AI-CoE's role is crucial in guiding LLMOps. It must lead the effort to standardize and optimize the utilization of LLMs, emphasizing their significance in providing ethical, efficient, and innovative applications. LLMOps can harness the potential of LLMs efficiently and responsibly while addressing the unique challenges associated with their resource management.

Resource management is not just a task; it's a critical aspect that impacts various dimensions of LLMOps. It directly influences cost, performance, scalability, reliability, security, and compliance. By implementing effective resource manage-ment strategies, contact centers can maximize the value derived from LLMs while minimizing risks and optimizing operational efficiency. The oversight of resources for LLMOps must involve implementing processes, tools, and best practices to monitor, manage, and optimize resources effectively.

Your AI-CoE must define clear resource allocation policies that align with contact center goals, budget constraints, and performance requirements. These policies should specify how resources are allocated, prioritized, and managed across different projects, teams, or departments. Analyze workload characteristics and resource requirements to determine the optimal placement of workloads across available infrastructure resources. Implement automation tools and scripts to streamline resource provisioning and scaling processes.

Implementing auto-scaling policies can automatically adjust resource capacity based on workload demand, optimizing resource utilization and cost efficiency. When making placement decisions, consider factors such as data locality, latency, cost, and performance. Utilize resource reservation mechanisms to guarantee access to critical resources for high-priority workloads or applications. Set resource quotas to prevent resource contention and ensure fair allocation of resources among different users or projects.

Utilize monitoring tools to track resource utilization metrics in real-time, including CPU usage, memory consumption, network bandwidth, and storage capacity. Set up alerts to notify stakeholders when resource usage exceeds predefined thresholds or when anomalies are detected. Track resource usage and associated costs closely to ensure alignment with budgetary constraints and cost optimization goals. Implement cost allocation tags, billing alerts, and budget management tools to monitor spending and identify cost-saving opportunities.

Implement governance policies and access controls to enforce compliance with regulatory requirements, security standards, and organizational policies. Ensure that access to sensitive resources is restricted based on role-based access controls (RBAC) and least privilege principles.

By following these best practices, your AI-CoE can effectively oversee resources for LLMOps, ensuring efficient resource utilization, cost optimization, performance reliability, and compliance with governance requirements.

It's About Analytics

Analytics in AI technologies is a foundational technology that enhances customer service and provides actionable insights. Contact centers increasingly use analytics to identify the behavior and demographics of customers to serve them more effectively. Analytics can forecast call volumes, predict customer behaviors, identify potential issues before they escalate, and personalize customer interactions. However, the effectiveness of these analytics depends significantly on the standards set for their deployment and use by your AI-CoE. While descriptive analytics provides a rear-view mirror perspective of "what" has happened, diagnostic analytics explores the "why" behind specific patterns or results.

Diagnostic analytics determines why certain days or times experience higher call volumes and identify factors such as promotional campaigns, product issues, or seasonal trends. For instance, if customer satisfaction noticeably drops, diagnostic analytics can help pinpoint whether it's due to longer waiting times, agent behaviors, or changes in service policies. Diagnostic analytics often involves drilling down into data, conducting data discovery, and performing inquiries to understand events' context and root causes.

Predictive analytics is forward-looking. It uses statistical models and forecasting techniques to understand the future. Predictive analytics relies on data, statistical algorithms, and machine learning techniques to identify the likelihood of future outcomes based on historical data. It answers questions like "What could happen?"

Prescriptive analytics is the next step after predictive analytics. It uses predictive models, AI, and machine learning to "recommend actions" for optimal results. This goes beyond descriptive, diagnostic, and predictive models by identifying patterns, forecasting outcomes, and recommending subsequent actions that can effectively influence those outcomes.

Each type of analytics plays a critical role in transforming data into actionable insights, with prescriptive analytics being particularly valuable for optimizing decision-making and improving CXs in contact centers.

The complexity and sensitivity of these tasks necessitate a structured approach to managing analytics of AI applications, which is where your AI-CoE becomes crucial. Start by clearly defining the objectives of using analytics in your contact center. Understand your specific goals, whether it's improving customer satisfaction, reducing wait times, or increasing sales through targeted recommendations. Analytics is all about the quality of data. Your AI-CoE must develop standards for data collection, storage, and processing to ensure ingestion into the models is accurate, consistent, and comprehensive.

To be effective, your AI-CoE must oversee the seamless integration of analytics with existing contact center systems, ensuring that predictive insights are seamlessly incorporated into your workflows. Your AI-CoE must define key performance indicators and metrics for evaluating the effectiveness of analytics. It must foster a culture of innovation and continuous improvement, staying abreast of the latest AI and data science developments to enhance analytical capabilities.

Your AI-CoE must also establish guidelines for ethical AI use, particularly in predictive and prescriptive analytics. This involves ensuring that models do not inadvertently perpetuate biases or lead to unfair outcomes for specific customer groups. Your AI-CoE must ensure analytic practices comply with relevant laws and regulations, such as data protection and privacy laws, which is vital for maintaining customer trust and avoiding legal issues.

Artificial Intelligence Center of Excellence

Talent & Expertise
- Leadership
- Project Management
- Strategic Partnerships
- Contact Center Domain Knowledge
- AI & ML Learning
- Data Science & Analytic Skills

6 Talent and Expertise

It's not all about talent. It's about leadership, management, partnerships, and being able to learn in order to improve.

Talent and expertise are a central area of focus for any AI-CoE to be successful, starting with a set of highly qualified and tenured experts who can sow the seeds for thought leadership and build a high-performing team to deliver the most significant impact through the following:

• Leadership
• Project management
• Strategic partnerships
• Contact center domain knowledge
• AI and machine learning skills
• Data science and analytical skills

The success of your AI-CoE hinges on being led by an intrapreneurial leader with a strategic vision that aligns AI initiatives with the overall business goals and customer service objectives. Your leadership must understand the technical aspects of AI and how these technologies can enhance CXs and operational efficiency. A key leadership responsibility is ensuring AI initiatives are scalable, sustainable, and capable of evolving with changing business needs and technological advancements. This involves continuous monitoring, evaluation, and optimization of AI applications to maximize their effectiveness and ROI. Moreover, leadership in your AI-CoE must prioritize ethical considerations and regulatory compliance. This includes addressing concerns related to data privacy, customer consent, and unbiased AI algorithms to ensure that AI applications respect customer privacy, adhere to ethical standards, and comply with industry-specific regulations.

Your AI-CoE involves overseeing project management for the development and deployment of AI solutions, from conceptualization to execution. This requires a thorough understanding of AI capabilities and limitations, as well as the specific needs and challenges of contact center operations. Project managers must adeptly coordinate cross-functional teams comprising AI experts, data scientists, IT professionals, and contact center staff.

There will be times when your AI-CoE will require external expertise from strategic partnerships to rapidly help the team gain real-world exposure and work on live projects. These partnerships will diversify risk and continually provide an outside-in context, creating standards and solutions that reflect industry trends.

Your team should be staffed by professionals with deep knowledge of AI-CoE team and machine learning, which is fundamental. A strong understanding of contact center dynamics, customer service processes, and industry-specific challenges is essential to address real-world issues and enhance your CX. They should possess skills in developing, training, and optimizing AI models, particularly those relevant to contact center operations, such as natural language processing, speech recognition, and predictive analytics. Your AI-CoE will also need experts in data science and analytics, which are crucial for interpreting and leveraging the vast amounts of data generated by contact centers. Skills include proficiency in statistical analysis, data visualization, and translating data insights into actionable business strategies.

30 The Role of Leadership

Leadership requires vision and a comprehensive understanding of both the potential and the challenges associated with integrating AI into customer service operations.

Leadership requires vision and a comprehensive understanding of both the potential and the challenges associated with integrating AI into customer service operations. It plays an important role in steering your organization towards innovation, ensuring that AI technologies are leveraged effectively to enhance customer experience, streamline operations, and maintain a competitive edge in the rapidly evolving digital landscape.

The core of your AI-CoE's mandate is developing and implementing a cohesive AI strategy that aligns with your broader business objectives of the contact center. This requires leaders who are not only well-versed in the technical aspects of AI but also possess a deep understanding of the contact center's operational dynamics and customer service ethos. By combining technical expertise with strategic business acumen, leaders can identify strategic areas within the contact center operations where AI can bring about transformative changes, such as automating routine inquiries, personalizing customer interactions, and predicting customer needs through advanced analytics.

Fostering a culture of innovation and continuous learning within the organization is also part of leadership. As AI technologies evolve, it is imperative for your AI-CoE to stay abreast of the latest developments and continuously assess their applicability to the contact center's needs. This necessitates a leadership approach that encourages experimentation, tolerates calculated risks, and views failures as opportunities for learning and growth. By promoting a culture that values agility and adaptability, leaders can ensure that the contact center remains at the forefront of AI-driven customer service innovation.

Moreover, the successful integration of AI into contact center operations hinges on collaboration and buy-in from various stakeholders, including frontline employees, IT staff, and executive leadership. Therefore, leaders within your AI-CoE must excel in change management and possess strong communication skills to articulate AI initiatives' vision, benefits, and implications. This involves highlighting the potential for enhanced efficiency and customer satisfaction and addressing concerns related to job displacement and the need for upskilling. Leaders must facilitate the transition and foster a sense of ownership and commitment across the organization by engaging stakeholders through open communication and involving them in the AI transformation journey.

Another aspect of leadership is emphasizing ethical considerations and responsible AI use. As AI technologies become more integral to contact center operations, leaders must ensure that AI tools are deployed to respect customer privacy, ensure data security, and promote fairness and transparency. This requires the establishment of ethical guidelines and governance structures that oversee AI deployments and address potential biases in AI algorithms.

Finally, leadership in an AI-CoE involves navigating the complex regulatory landscape that governs the use of AI in customer interactions. With data protection and privacy regulations becoming increasingly stringent, leaders must ensure that AI applications comply with relevant laws and industry standards. This requires a proactive approach to regulatory compliance, including regular audits of AI systems, ongoing training for staff on legal requirements, and collaboration with legal and regulatory experts. Leaders can mitigate legal risks and protect the organization's reputation by staying ahead of regulatory changes and ensuring compliance.

31 Manage Your Project

Project management is instrumental in ensuring the strategic alignment of AI initiatives with your contact center's broader business objectives.

Your AI-CoE is a specialized team aimed at promoting best practices and driving innovation in AI. It provides the necessary expertise and support to deploy AI solutions in contact centers effectively. Governance of project management from the AI-CoE entails being responsible for ensuring strategic alignment, executing projects, implementing governance mechanisms, managing knowledge, and driving continuous improvement. AI project oversight encompasses not only the traditional aspects of project management, such as scope, time, and cost accounting but also delves into the intricacies of contact center AI technologies. Your AI-CoE essentially bridges the gap between human representatives and AI technologies to create a balanced customer experience where customers do not feel neglected by automated systems but receive the personal touch they require from human employees. Furthermore, it involves adhering to relevant data privacy and usage laws, especially in highly regulated industries like healthcare.

At the outset, project management is instrumental in ensuring the strategic alignment of AI initiatives with your contact center's broader business objectives. Begin by working closely with stakeholders to define clear, measurable goals for what your AI-CoE aims to achieve, such as enhancing customer satisfaction, improving operational efficiency, or reducing costs. Facilitate this process by developing a comprehensive project charter, identifying KPIs, and establishing a roadmap that aligns AI projects with the organization's strategic vision. This step is foundational to secure executive sponsorship and cross-functional support.

Effective project execution is another aspect where project management methodologies come into play. Given the complexity and novelty of AI projects, adopting a flexible and adaptive project management approach, such as Agile or Scrum, can be particularly beneficial. These methodologies enable your AI-CoE to iterate quickly, adapt to changing requirements, and deliver tangible results in shorter cycles. Project managers need to direct the planning, scheduling, resource allocation, and risk management of AI projects. They must make designated teams remain focused, productive, and aligned with project objectives.

Governance is another project management responsibility within the AI-CoE. This involves setting clear policies, processes, and standards for developing and deploying AI solutions. Make sure your AI projects adhere to ethical guidelines, data privacy laws, and industry regulations, mitigating legal and reputational risks. Project management oversight also plays a key role in stakeholder management, communicating progress, identifying challenges, and documenting successes. Another role is setting up an effective project governance framework that includes decision-making bodies, such as steering committees, to provide oversight and direction for AI initiatives.

Continuous improvement is a fundamental principle in project management that is especially pertinent in the context of an AI-CoE. As AI advances, your AI-CoE needs to remain agile and responsive to new opportunities and challenges. This involves regularly reviewing and assessing the performance of AI projects against set KPIs, conducting post-implementation reviews to identify areas for improvement, and fostering a culture of experimentation and innovation. Overall, effective project management enables your AI initiatives to be strategically aligned, effectively executed and continuously improved.

32 Strategic Partnership

One of the most important advantages of strategic vendor partnerships is access to cutting-edge technology, specialized knowledge, and expertise.

Strategic vendor partnerships are indispensable to the success of contact center AI projects, especially when orchestrated by an AI-CoE. Partnerships with strategic vendors offer access to advanced technologies and drive innovation. It can provide collaborative and strategic alliances that leverage the strengths and capabilities of external vendors to accelerate innovation. The right strategic partner can allow your business to focus on its core competencies and support customization and integration efforts directed at improving your contact center's operations and CX. Many businesses rely on trusted networks of consultants, suppliers, and resellers to create these partnership ecosystems. Partnerships working in the context of AI are particularly important for small and medium-sized enterprises (SMEs) that have limited research and development funding.

The challenge to SMEs realizing their AI ambitions is simply a lack of skills. Much has been written about the shortage of AI professionals, such as machine learning engineers, data scientists, and project managers with experience in the field. Creating a partnership ecosystem includes technology and systems integrators, consultants, and software vendors, all of whom play a part in bringing the best together and filling gaps in skillsets, creating robust data strategies, or guiding organizational change.

One of the most important advantages of strategic vendor partnerships is access to cutting-edge technology, specialized knowledge, and expertise. Vendors that are industry leaders invest significantly in research and development to maintain market position. These vendors can offer your business state-of-the-art solutions beyond your organization's internal capabilities. By partnering with these vendors, your AI-CoE can leverage advanced technologies, including natural language processing, machine learning, and automation tools, to enhance customer service operations, reduce response times, and improve customer satisfaction.

Implementing AI in contact centers also involves various risks, including technological, regulatory, and operational challenges. Strategic vendor partnerships help mitigate these risks through shared responsibility. Vendors bring experience from working with multiple clients across industries, enabling them to foresee potential pitfalls and provide guidance on best practices and compliance issues. This collaborative approach helps navigate the complexities of AI implementation, ensuring that projects are completed successfully and comply with relevant regulations.

Strategic partnerships offer financial advantages. By leveraging economies of scale, vendors can provide cost-effective solutions that might be prohibitively expensive if developed in-house. Furthermore, these partnerships allow for scalability, enabling contact centers to adjust their AI capabilities as their needs evolve without significant additional investment in infrastructure or human resources. This flexibility is crucial for adapting to changing market demands and customer expectations.

It is important to follow certain best practices to maximize the benefits of collaborating with vendors on innovation. Clearly define your innovation goals, scope, and expectations, and communicate them to your vendors. Select vendors with the expertise, experience, and resources to support your innovation projects. Establish a formal agreement that outlines each party's roles, responsibilities, and deliverables, as well as the terms and conditions of the collaboration. Finally, set up regular meetings, updates, and feedback sessions to monitor the progress of the innovation projects and the success of your partnership.

33 Contact Center Domain Knowledge

Integration of contact center domain knowledge within your AI-CoE is not just beneficial but essential for crafting AI solutions that are impactful, innovative, and aligned with the core objectives of improving customer service.

Understanding the significance of merging contact center domain knowledge with AI requires a deep comprehension of both domains and how they work together for operational efficiency and technological innovation. By understanding the synergies of AI and contact center operations, your AI-CoE will have the transformative ability to integrate domain-specific insights with AI capabilities to meet and exceed the evolving expectations of both customers and businesses in the dynamic landscape of CX.

This integration of contact center domain knowledge within your AI-CoE is not just beneficial but essential for crafting AI solutions that are impactful, innovative, and aligned with the core objectives of improving customer service. Your contact center is the nerve center of customer interaction with your business. It is where customer queries are answered, support is provided, and feedback is collected. This communication hub is central to resolving issues and fostering strong relationships between your business and customers. Its domain has a comprehensive understanding of customer service workflows, communication channels (phone, email, chat, and social media), customer behavior, and the operational challenges inherent in managing these complex systems.

Your AI-CoE, on the other hand, should represent the pinnacle of your business's commitment to enhancing AI capabilities. It embodies a concentrated effort to innovate, standardize, and propagate AI technologies across various departments, ensuring that the benefits of AI are fully realized and aligned with the organization's strategic goals. This includes deploying AI to automate repetitive tasks, enhance decision-making with data analytics, and create new avenues for customer engagement through conversational AI and personalized experiences.

First, bringing contact center domain knowledge into your AI-CoE makes certain your AI initiatives are ingrained in the practical realities and specific challenges of customer service. By understanding the nuances of customer interactions and the operational complexities of contact centers, AI solutions can be tailored to address real-world problems. For instance, AI-powered chatbots can be designed to handle common queries with a level of understanding and responsiveness that closely mirrors human agents.

Secondly, incorporating contact center domain knowledge enhances the ability of AI systems to learn and adapt over time. AI models thrive on data; the varied interactions within contact centers allow these models to evolve. Understanding the context and significance of this data is necessary for AI systems to make meaningful improvements. Domain expertise supports training on data, directed by insights into customer behavior patterns, seasonal trends, and the impact of external factors on customer service needs.

Collaboration between domain experts and AI specialists also fosters a culture of continuous learning and improvement. As contact center professionals submit their practical insights and challenges, AI experts can offer technological solutions that might not have been apparent from within the contact center silo. This cross-pollination of ideas accelerates the pace of innovation and establishes that technological advancements are based on practical utility and customer value. As a result, customers are more likely to embrace AI-driven interactions if they feel understood and valued, with technology enhancing rather than detracting from the quality of service.

34 AI and ML Talent

AI and ML talent used in your AI-CoE should extend beyond the boundaries of technical development to strategic visioning and leadership.

Your AI-CoE cannot be a mere repository of tools and technologies. It needs to be a dedicated collection of talent and expertise to direct your business toward innovation in AI applications. AI and ML are not static; they continuously evolve and require specialized knowledge and skills to deploy them effectively. This is a full-time endeavor that requires a dedicated team of experts who are proficient not only in current methodologies but also in emerging techniques.

The convergence of AI and ML expertise enables a multidisciplinary approach to problem-solving. AI and ML applications often intersect with various domains, including data science, software engineering, ethics, and user experience design. By amalgamating talent from these diverse fields, your AI-CoE can create a holistic approach to AI development and deployment. This approach ensures solutions are technically sound, ethically responsible, and user-centric.

The way AI and ML talent is used in your AI-CoE should extend beyond the boundaries of technical development to strategic visioning and leadership.

AI and ML expertise in your AI-CoE should identify emerging AI trends and technologies. Their insights impact strategic decision-making, prioritizing investments in AI initiatives that align with long-term goals and market opportunities. This forward-looking perspective ensures that the organization remains at the cutting edge of AI advancements, ready to seize new opportunities.

In addition to motivating innovation and strategic planning, AI and ML talent in your AI-CoE helps with knowledge dissemination and skill development. The pace of change in AI technologies can create knowledge gaps, leaving your contact center staff behind without the skills necessary to leverage AI in their roles. The talent in your AI-CoE can bridge this gap through training programs, workshops, and mentorship, fostering a culture of continuous learning and AI literacy. By democratizing AI knowledge, your contact center can unlock the full potential of its workforce, encouraging cross-functional collaboration and the innovative application of AI across various business units.

The recruitment and retention of AI and ML talent becomes increasingly important with the success of your AI-CoE. The high demand for skilled AI professionals is competitive, making it essential for your business to create an environment that attracts and nurtures top talent. This includes offering opportunities for professional growth, fostering a culture of innovation, and providing access to cutting-edge technologies and projects. By investing in talent development, your business will be able to sustain the viability of your AI-CoE and its ability to innovate for the long term.

Overall, the presence of specialized AI and ML talent within an AI-CoE underscores your business's commitment to responsible AI use. Ethical considerations, data privacy, and security are paramount in AI applications. Experts in your AI-CoE must be tasked with developing and enforcing guidelines to be sure your AI solutions are developed and deployed ethically and transparently, respecting user privacy and societal norms. This commitment to responsible AI not only builds trust with customers and stakeholders but also positions your contact center organization as a leader in ethical AI practices.

35 Data Scientists and Analytical Skills

Analytical skills enable data scientists to critically evaluate the quality of data and the appropriateness of the analytical models they build.

As AI technology becomes more sophisticated, leveraging data effectively for decision-making, strategy formulation, and operational improvements becomes increasingly essential. Data science and analytical skills are not just supplementary talents within your AI-CoE but are central to its core mission. They are foundational to conceptualizing, developing, and refining intelligent systems. Designing and implementing AI solutions, particularly machine learning, depends on the ability to manage, process, and analyze large datasets, known as "big data."

Data scientists employ analytical skills to derive meaningful insights from complex datasets. They utilize statistical analysis, predictive modeling, and pattern recognition to transform raw data into actionable intelligence. These analytical endeavors are instrumental in identifying trends, testing hypotheses, and supporting decision-making processes.

Without a profound understanding of data science, your AI-CoE will struggle to build effective models that are the essence of AI functionality. Data science also enables the creation of bespoke AI solutions by identifying insights for customized models. Tailor-made approaches ensure that AI implementations are well-suited to the operational context of the organization, thereby maximizing their impact.

Analytical skills empower organizations to transition from intuition-based to empirical decision-making processes. Staffing your AI-CoE with professionals with strong analytical skills can help dissect complex problems, assess various metrics, and examine data trends to make informed decisions. This empirical approach to decision-making can significantly enhance AI solutions developed by your contact center, ensuring they are grounded in real-world evidence and capable of delivering tangible business outcomes.

Personalization is essential to customer satisfaction in any business, and the analytical skills in your AI-CoE are necessary to understand customer behavior, preferences, and feedback. By analyzing customer data, AI-powered data science can identify patterns in behavior and preferences from previous interactions to tailor experiences, recommendations, and services to individual customers. For instance, data analytics can enable predictive personalization, where AI anticipates a customer's question or issue based on historical data and provides customized responses or recommendations. Additionally, sentiment analysis can assess the tone and emotion in a customer's communication, enabling agents to adjust their approach accordingly. This individualized attention enhances customer satisfaction and loyalty.

AI necessitates constant innovation and improvement. Data science and analytics facilitate the iterative process of model training and refinement. By analyzing the performance of AI systems and identifying areas for enhancement, data scientists can apply their expertise to innovate and continually evolve AI solutions. This iterative cycle is essential to maintaining the relevancy and efficacy of AI applications. Your AI-CoE must also be responsive to market changes. Data science and analytics must provide the agility to swiftly adapt AI strategies in response to new trends, emerging data sources, and competitive pressures. The insights from ongoing data analysis enable your AI-CoE to pivot and innovate, maintaining a competitive edge in the marketplace.

Your AI-CoEs must operate at the intersection of multiple domains, requiring a collaborative approach. To deliver comprehensive AI solutions, data science and analytics professionals must work alongside other experts, such as software engineers, business analysts, and domain specialists. Their ability to interpret data and extract insights is critical in bridging the gap between technical and business stakeholders, ensuring that AI initiatives are aligned with organizational objectives.

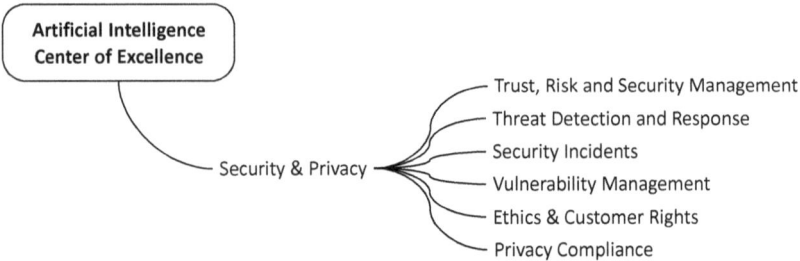

Artificial Intelligence Center of Excellence

Security & Privacy
- Trust, Risk and Security Management
- Threat Detection and Response
- Security Incidents
- Vulnerability Management
- Ethics & Customer Rights
- Privacy Compliance

7 Security and Privacy

The domain of security and privacy requires an approach to enhance AI capabilities in several key areas:

• Threat, risk, and security detection
• Threat, detection, and response
• Vulnerability management
• Ethics and customer rights
• Privacy compliance

This approach requires your AI-CoE to fortify your business defenses against evolving cyber threats while ensuring adherence to your contact center's ethical guidelines and privacy regulations.

For threat detection, your AI-CoE needs to elevate the efficiency and effectiveness of threat detection by leveraging advanced AI algorithms to analyze vast datasets more rapidly than human analysts. AI can identify potential threats that traditional methods may not detect. AI/ML models can recognize patterns that indicate cyberattacks, including zero-day threats. These models learn continuously from new data, enabling them to improve their detection capabilities over time.

For incident response, your AI-CoE can contribute by enabling quicker and more accurate responses to security incidents. AI-driven systems automate the initial stages of an incident response, such as alert triaging and initial analysis, freeing up valuable time for human experts to focus on more complex tasks. Additionally, AI can assist in predicting the trajectory of an ongoing attack, allowing organizations to mitigate its impact preemptively.

Your AI-CoE can enhance vulnerability management through the automation of detection and prioritization of system vulnerabilities. Traditional vulnerability management can be labor-intensive and prone to human error, but AI algorithms can continuously scan an organization's digital infrastructure for vulnerabilities. More importantly, AI can prioritize these vulnerabilities based on potential impact, enabling organizations to allocate their resources more effectively.

For risk management, your AI-CoE must oversee advanced analytical capabilities to identify and assess potential risks. AI models can process and analyze large volumes of data to forecast potential security risks and their probable impacts on the organization. This predictive analysis is crucial for developing effective risk mitigation strategies. Furthermore, AI-CoEs facilitate dynamic risk management, allowing organizations to adapt their security posture in response to changing threat landscapes.

Your AI-CoE needs to be accountable for ensuring that AI deployment aligns with ethical standards for customer rights. It must also develop and enforce ethical guidelines for AI usage, ensuring that AI-driven decisions are transparent, fair, and non-discriminatory. This includes addressing biases in AI algorithms and safeguarding customer data against misuse. By upholding ethical principles, AI-CoEs help maintain public trust in AI technologies, indispensable for their widespread acceptance and utilization.

For privacy compliance, particularly as data protection regulations are becoming increasingly stringent, your AI-CoE must make sure AI systems are compliant with regulations. It must ensure personal data is processed lawfully, transparently, and securely. This involves developing AI solutions that incorporate privacy by design, ensuring that data privacy is an integral part of the system from the outset. Additionally, your AI-CoE needs to automate certain aspects of privacy compliance, such as data subject access requests and data protection impact assessments, making it easier for organizations to adhere to regulatory requirements.

36 Trust, Risk, and Security Management

As AI is increasingly adopted, it is imperative to strike a balance between harnessing its power and ensuring its responsible and secure use.

AI Trust, Risk, and Security Management (AI TRISM) form a framework for your AI-CoE that encompasses AI model trustworthiness, fairness, reliability, robustness, efficiency, and data protection. As your business increasingly adopts AI technologies, it becomes important to strike a balance between harnessing the power of AI and ensuring responsible and secure use.

AI trust management is central to AI governance and essential for AI acceptance and effectiveness. To build trust, transparency is paramount. Stakeholders, including customers, employees, and regulatory bodies, should have visibility into how AI models make decisions, referred to as Explainable AI (XAI). Your AI-CoE needs to invest in XAI technologies and methodologies that enable them to understand and communicate how AI models arrive at specific conclusions.

Enforcing fairness in AI models is another cornerstone of trust management. Bias in AI can lead to unfair or discriminatory outcomes. In contact centers, where AI often interacts with customers, bias can have profound consequences. To address this, your AI-CoE must implement mechanisms to detect and mitigate bias in AI models with regular fairness audits to identify and rectify any disparities in the treatment of different groups.

Another facet of trust management involves accountability. Your AI-CoE must define clear lines of responsibility for AI systems. Knowing who is responsible for developing, maintaining, and overseeing AI models is critical. This ensures accountability for AI performance and facilitates swift action in case of errors or issues.

Data privacy is important to trust management, particularly in contact centers that handle sensitive customer information. AI models must be developed and deployed with robust data privacy measures. Compliance with regulations is non-negotiable. This involves data anonymization, encryption, and stringent access controls to safeguard customer data.

Recognize that AI introduces new risks and must be proactively managed. Risks may include system failures, unexpected behaviors, or inaccurate predictions. For contact centers relying on AI-driven chatbots or voice recognition systems, a system failure can disrupt customer interactions and harm your CX. Risk mitigation strategies should be in place, including robust testing, monitoring, and failover mechanisms.

Ethical risks are a growing concern in AI models and can inadvertently perpetuate biases present in the training data. In the context of contact centers, this could result in discriminatory treatment of customers based on factors like race or gender. Your AI-CoE must actively identify and mitigate ethical risks associated with your AI models. This may involve retraining models with more diverse and representative datasets or employing bias detection and correction tools. AI models and the data they process are valuable assets that cybercriminals can target. Robust cybersecurity measures are essential to protect AI models and the sensitive data they handle. This includes measures like encryption, intrusion detection systems, and secure deployment of AI in isolated environments.

37 Threat Detection and Response

Establishing and enforcing robust security protocols and ethical guidelines requires continuously identifying potential threats, including data breaches, biased algorithms, or misuse of AI technologies.

In threat detection and response, your AI CoE's primary role is establishing and enforcing robust security protocols and ethical guidelines. This involves continuously identifying potential threats, including data breaches, biased algorithms, or misuse of AI technologies.

AI can predict potential future threats and vulnerabilities by analyzing historical data and security events. This proactive approach allows organizations to address security gaps before malicious actors exploit them. Your AI-CoE must anticipate new threats emerging as AI technologies evolve by staying abreast of the latest developments in AI and cybersecurity. It is responsible for harnessing AI capabilities, such as advanced algorithms, machine learning, and deep learning techniques, to analyze vast amounts of data and identify patterns that human analysts may overlook. By doing so, your AI-CoE can proactively develop strategies to mitigate these threats.

Overseeing the implementation of AI solutions across your organization will make sure your AI applications are developed and deployed in a manner that adheres to established security and ethical standards. Your AI-CoE can facilitate this by providing expertise and guidance to different teams within the organization, helping them to understand the potential risks associated with AI technologies and how to mitigate them.

In addition to setting standards and guidelines, your AI-CoE must develop and maintain the tools and technologies to detect and respond to AI-related threats. This could involve implementing advanced cybersecurity measures, such as intrusion detection systems specifically designed to protect AI systems. Your AI-CoE might also develop or adopt tools for monitoring the performance and behavior of AI applications, ensuring that they are functioning as intended and identifying any anomalies that could indicate a security or ethical issue.

Your AI-CoE must also play a role in response planning and management. In the event of a security breach or ethical violation, your AI-CoE should have a clear plan for how to respond. This plan should include steps for containing and mitigating the breach, investigating its cause, and implementing measures to prevent similar incidents in the future. Your AI-CoE should also be prepared to work with external stakeholders, such as regulatory bodies and law enforcement, as part of the response effort.

Another aspect of your AI-CoE's role is fostering a culture of responsibility and accountability within the organization. This involves ensuring all stakeholders, from executives to frontline employees, understand their role in maintaining AI systems' security and ethical integrity. Your AI-CoE can promote this culture through regular communication, reporting on AI security and ethics status, and recognizing individuals or teams contributing to these efforts.

By leveraging advanced AI capabilities, your AI-CoE enables organizations to detect and respond to security threats proactively, automate incident response processes, and empower security analysts with advanced tools for threat hunting. This ultimately enhances an organization's cybersecurity posture and enables your AI-CoE to stay ahead of evolving cyber threats.

38 Manage Security Incidents

AI enables early threat detection to identify potential security breaches or privacy concerns before they escalate into full-blown incidents.

Integrating advanced AI technologies with cyber-security practices can improve your business's ability to prevent, detect, respond to, and recover from incidents. AI technologies, such as ML algorithms and NLP, are adept at analyzing vast amounts of data at speeds unattainable by human operators. AI enables early threat detection to identify potential security breaches or privacy concerns before they escalate into full-blown incidents. For example, by continuously monitoring network traffic and user behavior, AI systems can spot anomalies that may indicate a cyberattack, such as unusual access patterns or malicious code, with greater accuracy and efficiency than traditional methods.

In the event of a security breach or data privacy incident, time is of the essence. Your AI-CoE should govern the AI-driven tools to automate aspects of the response process, from categorizing the type of incident and assessing its severity to recommending or even implementing remedial actions. These speed up the response time and help allocate human resources more effectively, ensuring that cybersecurity professionals focus on the most critical aspects of an incident. Furthermore, AI can facilitate dynamic risk assessment, adjusting the organization's security posture in real-time based on evolving threats, thereby enhancing resilience against future incidents.

Privacy preservation is another critical area where your AI-CoEs can significantly contribute. With regulations such as the General Data Protection Regulation (GDPR) imposing strict rules on data handling, organizations must ensure that personal information is protected against unauthorized access and breaches. Your AI-CoE must implement and oversee the deployment of AI models designed with privacy in mind, such as those utilizing techniques like differential privacy or federated learning. These models can analyze sensitive data without exposing individual details, thus maintaining privacy while deriving valuable insights. Additionally, AI can assist in compliance monitoring, automatically scanning for and rectifying potential privacy violations, further embedding privacy considerations into the organization's operational fabric.

The effectiveness of your AI-CoE in incident management hinges on its ability to navigate the ethical and technical challenges associated with AI. This includes ensuring AI systems' transparency, fairness, and accountability; addressing biases in AI models that could undermine security efforts; and safeguarding against adversarial AI tactics. To this end, your AI-CoE must invest in continuous research and development, collaborate with external experts and institutions, and adhere to ethical guidelines and standards. Such measures are vital for maintaining trust in AI applications and, by extension, the organization's overall security and privacy framework.

Your AI-CoE has the strategic role of promoting a culture of security and privacy awareness within the organization. This can be achieved by leading educational initiatives and training programs for disseminating knowledge about cybersecurity and best practices for ensuring data privacy. Regular simulations, updates to AI models, and cross-training on AI systems enhance the ability of human oversight. Both AI algorithms and human operators benefit from ongoing education and training to stay abreast of evolving incident scenarios. This holistic approach equips your contact center with the necessary skills to contribute to incident prevention. It promotes a shared responsibility model, where security and privacy are seen as collective endeavors.

39 Vulnerability Management

Vulnerability management requires a strategic approach encompassing risk management, continuous monitoring, education, compliance, and collaborative improvement. It involves developing and implementing a set of practices designed to identify, assess, mitigate, and protect against vulnerabilities in AI systems. Given the complexity of AI systems and their integration into critical business processes, a proactive and comprehensive approach to vulnerability management is not just an option; it is a necessity for maintaining operational integrity, security, and trust.

Governance of vulnerability management in your AI-CoE begins with establishing a framework. A governance framework delineates roles and responsibilities, integrates with the broader organizational cybersecurity policies, and ensures alignment with overall business objectives. It articulates the decision-making processes, sets out the oversight mechanisms for vulnerability management activities, and defines the escalation procedures during a security incident. Your framework must include a strategy for risk assessment to manage vulnerabilities effectively. Risk assessment identifies and prioritizes potential risks and threats, allowing for more targeted and efficient vulnerability management.

Your AI-CoE needs to perform risk assessments at regular intervals to identify potential vulnerabilities in your AI systems. These assessments should cover ML models, data, and your underlying infrastructure. Based on the assessment results, a risk management plan should be developed outlining how to address high-risk vulnerabilities through remediation, mitigation, or acceptance, depending on your business's risk appetite.

Identifying and tracking vulnerabilities is a dynamic process that necessitates continuous vigilance. Your AI-CoE needs to evaluate automation tools that can scan and monitor the AI environment for known vulnerabilities. At the same time, reviews by your manual expert can provide insight into more nuanced or emerging threats. Ideally integrated into the AI-CoE's IT management suite, a centralized tracking system is essential for documenting and following up on identified vulnerabilities to ensure they are addressed within appropriate timeframes.

Identifying and tracking vulnerabilities is an ongoing process that requires constant attention. Automation tools can scan and monitor the AI environment for known vulnerabilities, while manual expert reviews can provide insight into more complex or emerging threats. A centralized tracking system, ideally integrated into the AI-CoE's IT management suite, is essential to document and follow through on identified vulnerabilities and ensure they are addressed within appropriate timeframes.

Patch management is also a significant component in governing potential system vulnerabilities. A patch management policy details the process for developing, testing, approving, and deploying patches and updates to AI systems. This policy needs to address the unique challenges AI systems pose, such as retraining or recalibrating models following a patch to prevent degradation of the system's performance or accuracy.

A comprehensive incident response plan is equally essential to vulnerability management. This plan must include clear procedures and responsibilities for detecting, reporting, and responding to security incidents within the AI environment. The plan should cover all phases of incident response: preparation, identification, containment, eradication, recovery, and post-incident analysis.

Lastly, continuously monitoring your AI environment is essential to identifying and responding to threats in real time. Utilizing advanced monitoring and analytics tools will measure the effectiveness of the vulnerability management program and detect patterns that might indicate a security breach or vulnerability exploitation attempt.

40

Ethics and Customer Rights

As AI becomes more prominent in your contact center, your AI-CoE should establish ethical dimensions and customer rights, particularly regarding security.

AI ethics and customer rights is a collection of moral principles that shape the responsible development, deployment, and use of AI technologies. As AI becomes more prominent in your contact center, your AI-CoE should establish ethical dimensions and customer rights, particularly regarding security. This involves a holistic approach to designing, deploying, and managing AI systems to ensure they are secure, transparent, and respectful of customer privacy and rights. Establishing this approach involves adhering to ethical guidelines, ensuring transparency, mitigating biases, prioritizing data security, maintaining truthful output, and addressing concerns related to job displacement. By incorporating these principles into the development and deployment of AI technologies, your business can uphold customer rights, enhance security measures, and build trust with their customers while leveraging the benefits of artificial intelligence your contact center operations.

Isaac Asimov foresaw the potential dangers of autonomous AI agents in his 1942 short story "Runaround," included in his 1950 collection, *I, Robot*. Asimov's "Three Laws of Robotics" were a means to limit those risks. In his code of ethics, the first law forbids robots from actively harming humans or allowing harm to come to humans by refusing to act. The second law orders robots to obey humans unless the orders are not in accordance with the first law. The third law orders robots to protect themselves insofar as doing so is in accordance with the first two laws. As businesses turn to AI technology to enhance CX, it is crucial for your AI-CoE to develop your own AI ethics guidelines to confirm your AI systems are deployed responsibly. This includes addressing issues such as bias, transparency, data privacy, and fairness in customer interactions.

Ethics and customer rights are already global concerns, with governments addressing the issue through legislation. The European Union Artificial Intelligence Act (AI Act) was proposed by the EU Commission on 21 April 2021, and passed on 13 March 2024. The AI Act establishes a common regulatory and legal framework for artificial intelligence. It follows a risk-based approach, classifying AI systems into five categories of systems: prohibited, high-risk, low-risk, minimal-risk, and general purpose. As the risks increase, so do corrective measures. The highest level of risk triggers a ban on AI systems. For less risky AI systems, the focus is on transparency obligations to ensure users know they are interacting with an AI system, not a human being.

AI ethics are also being encouraged domestically. In October 2022, the White House Office of Science and Technology Policy (OSTP) released the Blueprint for an AI Bill of Rights as an ethical framework for using AI in the U.S. This Blueprint contains five principles that the OSTP believes every American should be entitled to, including:

• Safe and effective systems: AI systems should be safe and effective.
• Algorithmic discrimination protections: AI algorithms should prioritize fairness and equity and must not contribute to discrimination in any way.
• Data privacy: AI systems should include built-in protections to safeguard data.
• Notice and explanation: Businesses should inform users whenever they use an AI system.
• Human alternatives, consideration, and fallback: Users should be able to opt out of using an AI system and receive help from a person when applicable.

41

Privacy and Compliance

Your AI-CoE must enforce a 'privacy by design and default' framework embedded into the development phase of any AI solution.

The foundation of privacy governance is a mainstay of AI that includes strategic development, best practices, and oversight of technological initiatives that are integral to AI. To build this foundation, your AI-CoE needs to orchestrate an approach to privacy, with comprehensive privacy grounded in transparency, accountability, and user empowerment. These policies dictate how personal and sensitive data should be handled throughout its lifecycle to certify your AI initiatives comply with data protection laws such as the General Data Protection Regulation (GDPR) or the Health Insurance Portability and Accountability Act (HIPAA).

To operationalize these principles, your AI-CoE must enforce a 'privacy by design and default' framework. Privacy considerations should be embedded into the development phase of any AI solution, with the default settings to maintain the highest standard of privacy protection. Create a roadmap that guides the design of AI applications and supporting infrastructure through the system development life cycle, from the initial design stages to deployment and beyond. Define testing metrics for verification and feedback to measure compliance performance. Interpret IT privacy and security regulatory standards, identifying gaps and assisting in closing them. Then, provide oversight to enforce policies, procedures, and standards.

Privacy governance is dynamic, and your business must be agile enough to respond to evolving legal requirements and technological advancements. Implement regular privacy impact assessments to provide structured evaluations of your AI projects. These assessments identify potential privacy breaches and mitigate any identified risks before they materialize through metrics and independent testing for verification and feedback.

A proactive stance on privacy extends to consent management. Your AI-CoE must implement thorough mechanisms for obtaining, documenting, and managing user consent and comply with legal standards. This includes:

- Clearly defined policies regarding data collection, processing, and storage to align with privacy laws and articulate the organization's commitment to data protection
- Collecting only the data necessary for the specified purpose to reduce privacy risks
- Obtaining, managing, and documenting consent management for user data processing activities
- Facilitating customer rights to access, rectify, delete, or port personal data

Your AI-CoE's commitment to compliance governance is to embed legal and ethical standards into AI systems and ensure your AI deployments are by industry standards, legal regulations, and ethical guidelines. This involves:

- Keeping abreast of relevant laws and regulations and ensuring AI systems are updated to remain compliant as these requirements evolve
- Regular audits to assess compliance and findings are appropriately reported to stakeholders and regulatory bodies
- Educating employees about compliance requirements related to AI use, including privacy laws, ethical considerations, and security practices
- Managing risk processes such as risk data information, data protection testing, investigation in case of breaches, root cause analysis, and remediation management

Governance over privacy and compliance is inherently collaborative. To create an integrated approach, your AI-CoE must interface with legal, IT, HR, and operations. This collaboration guarantees that privacy and compliance are not treated in isolation but are recognized as cross-functional responsibilities. Ultimately, continuous improvement is the ethos that drives your AI-CoE's governance activities, guaranteeing that privacy and compliance governance are not static but evolve to meet your business's challenges.

8 Conclusion

Your AI-CoE must be more than an entity; it is an idea, a commitment to excellence in AI. As your business addresses the spectrum of challenges and opportunities to innovate and transform AI, an AI-CoE becomes a hub to guide adoption and integration. It encapsulates the collective wisdom of strategic alignment, leverages cutting-edge technology, establishes sound governance, curates and protects data, extracts insights through advanced analytics, cultivates talent, and guards security and privacy.

Strategic alignment ensures that every AI initiative is not an isolated experiment but an influential part of your contact center strategy to drive value and create a competitive edge. To remain in concert with business stakeholders, your AI-CoE must provide oversight of innovative applications that resonate with your core objectives, enabling the advancement toward your end-state vision.

The deployment of technology under the aegis of the AI-CoE is not merely about adoption but about discerning selection and integration. The tools and platforms must be technologically advanced and align with your contact center's infrastructure and future scalability needs. This technological stewardship is critical to building a resilient and adaptable AI ecosystem that can evolve with the enterprise and the ever-changing technological landscape.

Governance is about establishing frameworks and policies that guide AI's ethical and responsible use. It involves setting standards for model development, decision-making processes, and metrics to evaluate AI performance and impact. It must also set policies that promote trust and accountability, warrant compliance with regulations, and weigh every decision for its impact on your contact center and customers.

Data is the lifeblood of AI, and your AI-CoE must champion the data life cycle from collection to disposal. Your AI-CoE must create data strategies, manage data repositories, and establish data governance practices to ensure that data is consistent, secure, and used in accordance with data protection laws. It must advocate for data integrity, recognizing that the strength of AI predictions and prescriptions is directly proportional to data quality.

Analytics is central to your AI-CoE, transforming data into actionable intelligence. With sophisticated analytical models and algorithms, your AI-CoE enables your contact center to discern patterns and glean previously imperceptible insights, thus supporting informed decision-making and innovation.

Nurturing talent within the AI-CoE cannot be overstated. It is the crucible where expertise is fostered and innovation thrives. By attracting, training, and retaining the brightest minds in AI, your AI-CoE can become a hub of creativity and a repository of institutional knowledge.

Your AI-CoE must provide safeguards from threats by establishing cybersecurity protocols, monitoring systems for potential breaches, and developing contingency plans to mitigate the impact of security incidents.

In conclusion, your AI-CoE must be an integral part of your contact center, which aims to be at the forefront of AI adoption. It centralizes expertise and leadership, facilitates strategic AI initiatives, and ensures that AI is utilized ethically, compliantly, and aligned with the organization's mission and values. Through strategic alignment, technology, governance, data, analytics, talent, security, and privacy, your AI-CoE will drive AI initiatives and embed a culture of innovation and responsible AI use in your contact center.

42

What Are Your Rules?

For leaders who have deployed AI or are planning AI, there are additional rules to be defined.

This book is the third in a trilogy of books to guide readers on using, planning, and managing AI in contact centers. *42 Rules for Using AI in Contact Centers* describes the possibilities of using AI. *42 Rules for Planning AI in Your Contact Center* defines the steps to plan AI. The rules in this book offer a blueprint for establishing and nurturing an AI-CoE for your contact center that is effective, practical, productive, and principled. The intent is for readers to carry forward the insights from these pages to build their own AI-CoEs and drive AI innovation to benefit their contact centers and customers.

While the trilogy is the culmination of the insights of numerous articles and whitepapers on what an artificial intelligence center of excellence should be for contact centers, it is an incomplete vision. The ultimate vision should be determined by the reader, as well as by the unique requirements of their contact center and the needs of their respective customers.

My rules are meant to identify discussion points to govern AI. Each rule contains a high-level description of an AI-CoE component and subcomponents. They are intended to explain that AI initiatives are not just to advance technology but are a necessity for your contact center to harness and manage AI through a center of excellence that provides a structured approach to navigating the complexities of its integration.

AI is still a new technology. The integration and methodologies for AI in contact centers are emerging unprecedentedly. Regardless of size, organizations may find themselves at a crossroads, needing to adopt AI to remain competitive while also facing the challenge of doing so ethically and effectively. My rules present an overview of the AI ecosystem and guide leaders through leveraging AI technologies to achieve business objectives. It is my hope that it demystifies AI, making it more accessible to non-technical stakeholders, and offers an inclusive environment where a combination of technological and business insights drives AI initiatives.

The question of governance and ethics in AI will be of ongoing importance. An AI-CoE is not merely a hub for innovation but also a custodian of ethical AI practices. By detailing how to establish governance structures and ethical guidelines within the CoE, my rules attempt to identify a broader discourse on responsible AI. It educates organizations on the importance of ethical considerations, such as fairness, transparency, and accountability, ensuring that AI solutions are developed and deployed in a manner that respects societal values and norms.

Moreover, my rules elucidate an AI-CoE's role in driving organizational transformation. An AI-CoE acts as a catalyst for change, enabling businesses to adapt to AI. It provides leaders with a roadmap for integrating AI into their corporate strategy, culture, and operations. My rules highlight an AI-CoE's role in fostering a culture of innovation and breaking down silos, thereby ensuring that the benefits of AI are realized across the organization.

For leaders who have deployed AI or are planning AI, there are additional rules to be defined. So, what are your rules?

Three Letter Acronyms & Lexicon

Artificial intelligence and the contact center industry are full of three or more letter acronyms (TLAs) and jargon that can overwhelm even the most knowledgeable individuals. This appendix summarizes TLAs and terms used in this book, but by no means is it a complete lexicon of all the terms used in either contact centers or the AI industry.

ABAC	Attribute-Based Access Control is a security model that determines access permissions based on attributes associated with users, resources, actions, and environmental conditions. ABAC allows for more fine-grained, context-sensitive access decisions than role-based access control, which relies solely on a user's role to grant or deny permissions. This approach enhances security and operational efficiency in contact centers by ensuring that agents, administrators, and other staff have access only to the specific information and functions necessary for their roles and tasks.

Accelerated Computing	Advances generative AI by significantly reducing training and inference times using specialized hardware such as graphics processing units (GPUs) and tensor processing units (TPUs). Generative models can leverage parallel processing to accelerate computations and efficiently handle complex tasks. This acceleration enables researchers and practitioners to train larger models, explore more sophisticated architectures, and achieve higher-quality outputs in various generative AI applications, ranging from image synthesis to natural language generation.
ACW	After-Call Work is a set of necessary tasks that need to be completed after an agent interacts with the customer. These include updating the system, logging the reason for contact and outcome, updating colleagues, and scheduling follow-up actions. ACW varies across different customer queries and resolution requests. There is no predetermined yardstick for the duration of ACW. However, it is a metric that impacts average handle time (AHT) and must be closely tracked.
ADC	Automatic Call Distributor is a telephony system that manages incoming calls and routes them to specific agents or departments within a contact center based on predefined rules and criteria. These criteria can include agent availability, skill sets, or other factors that contribute to efficient call handling and improved customer service. An ACD optimizes resource allocation in the contact center, thereby reducing customer wait times and enhancing the overall efficiency and effectiveness of the operation.

Agent Assist	Refers to the utilization of artificial intelligence and other advanced technologies to support human customer service agents in their tasks, including call handling, problem-solving, and customer engagement. These systems typically offer real-time guidance, automated responses, and contextual information to help agents efficiently address customer inquiries and concerns. The ultimate goal is to enhance the customer experience by expediting resolution times, improving accuracy, and allowing human agents to focus on more complex or emotionally nuanced interactions.
AGI	Artificial General Intelligence or strong AI represents a theoretical form of AI that proposes solving any number of hypothetical tasks using generalized human cognitive abilities. In theory, AGI will be able to understand, learn, and apply knowledge across a wide range of functions, similar to humans. The ultimate goal of AGI is to replicate the broad range of human cognitive abilities called common-sense reasoning. AGI research is still evolving, and researchers are divided on the approach and timelines to bring it to reality.
Agile Methodology	An iterative and incremental approach to software development that emphasizes flexibility, collaboration, and customer feedback. It involves breaking down projects into smaller, manageable tasks called user stories, which are prioritized and completed in short development cycles known as sprints, typically lasting one to four weeks. Agile teams regularly review and adjust their plans based on feedback, allowing them to respond quickly to changes and deliver value to customers faster. See Waterfall Methodology.
AHT	Average Handle Time is a contact center metric used to measure the average duration of one transaction. It usually starts when the customer begins the interaction and covers hold time, talk time, and any other related tasks during the conversation.

AI	Artificial Intelligence refers to developing and implementing computer systems that can perform tasks typically requiring human intelligence, such as learning, problem-solving, and decision-making. Examples of AI include understanding human text and speech and detecting and translating languages.
AI Act	A regulatory framework proposed by the European Union to govern the development, deployment, and use of artificial intelligence systems within the EU. It aims to ensure AI's ethical and trustworthy use while promoting innovation and competitiveness. The Act introduces requirements for high-risk AI systems, including transparency, accountability, and human oversight. It establishes a comprehensive framework for AI governance and market surveillance to protect the rights and safety of individuals.
AI Domain	The range of AI technologies specifically tailored for customer service environments, including but not limited to natural language processing (NLP), machine learning algorithms, and automated decision-making systems. These technologies are designed to assist or automate various functions within the contact center, such as customer engagement, routing, and analytics.
AI Domain Description	Detailed explanations or specifications of the specific AI domains that refer to particular areas or fields of knowledge where AI techniques and technologies are applied to solve certain problems or address specific challenges. Domain descriptions provide a deeper understanding of the AI technologies, their applications, and how they fit into your contact center setup. Domain descriptions could include information about the capabilities of AI technologies within the chosen domain, their benefits, potential use cases, and considerations for implementation, as well as define the scope and context of your AI application.

AI Project Lifecycle	Outlines the sequential phases involved in planning, developing, deploying, and maintaining AI technologies within a customer service environment. The lifecycle typically starts with problem identification and requirement gathering, followed by the design, development, and testing of the AI solutions, and culminates in deployment, monitoring, and ongoing optimization. Each phase has its own set of methodologies, best practices, and metrics for success, ensuring that the AI initiative aligns with the contact center's broader operational and business objectives.
AI TRISM	The governance framework that encompasses Trust, Risk, Isolation, Security, and Management of AI systems. In a CoE, AI TRISM ensures that the deployment and operation of AI technologies align with organizational values and compliance requirements while managing potential risks. It focuses on building reliable AI systems that are secure and isolated from unauthorized access, underpinned by a comprehensive risk management strategy and trustworthiness in AI outcomes.
Algorithm	Algorithms are a set of rules a computer follows while executing operations. Algorithms tell a computer how to act in various situations. Combining multiple algorithms allows applications to perform more sophisticated tasks without human intervention. For example, a chatbot can use algorithms to suggest products based on a shopper's purchase history or route customers to a specific human agent whose specialty best matches the incoming question.
ANN	Artificial Neural Networks are biologically inspired computational networks that simulate the human brain processes. ANNs consist of interconnected nodes, called artificial neurons or units, organized into layers which include an input layer, one or more hidden layers, and an output layer. ANNs can learn from data by adjusting the weights and biases of the connections between neurons, enabling them to process complex information, recognize patterns, and make predictions or decisions. ANNs have been successfully applied to various tasks, such as image recognition, natural language processing, and time series prediction.

Anonymization   The process of irreversibly transforming or
                  removing personally identifiable information from
                  a dataset so that individuals cannot be readily
                  identified. Unlike redaction or data masking, which
                  may allow for the possibility of re-identification,
                  anonymization aims to eliminate that risk entirely.
                  In the context of contact centers, anonymization is
                  often used to protect customer privacy and ensure
                  compliance with data protection regulations while
                  still allowing for the analysis and utilization of the
                  data for operational improvements and insights.
                  See Redaction.

AR   Augmented Reality is an interactive experience
                  that enhances the real world with computer-
                  generated perceptual information. Using software,
                  apps, and hardware such as AR glasses,
                  augmented reality overlays digital content onto
                  real-life environments and objects.

ASA   Average Speed of Answer is a KPI commonly used
                  in contact centers to measure the average time it
                  takes for an agent to answer incoming calls. The
                  metric is usually calculated by dividing the total
                  wait time for answered calls by the total number of
                  answered calls within a specific time period. ASA
                  serves as an important gauge for assessing the
                  efficiency and responsiveness of a contact center,
                  with shorter ASA times generally indicating better
                  customer service and resource allocation.

ASR   Automated Speech Recognition, also known as
                  speech recognition, computer speech recognition,
                  or speech-to-text, enables a program to process
                  human speech into a written format. ASR uses
                  algorithms and models to analyze and transcribe
                  audio signals into textual representations. ASR
                  applications are in various domains, including
                  voice assistants, transcription services, call
                  centers, and language processing, enabling
                  efficient and accurate conversion of spoken words
                  into text data.

Avatar	A digital representation or embodiment of an entity, often a person or a character that interacts with users in virtual environments or through digital platforms. Avatars can be visual representations in the form of animated characters or graphical icons, or they can be voice-based representations in the case of virtual assistants or chatbots. They are designed to simulate human-like behavior and engage in conversational interactions, providing personalized and interactive experiences for users.
BERT	Google BERT, or Bidirectional Encoder Representations from Transformers, is a natural language processing (NLP) model developed by Google. It employs a transformer architecture that considers both the left and right context of a word, enabling it to better understand the meaning of words within sentences. By capturing contextual relationships more effectively, BERT has significantly improved the performance of various NLP tasks, such as question answering, sentiment analysis, and language understanding.
Bias	When an algorithm shows prejudice in favor of or against one thing, person, or group compared with another, usually in a way considered unfair. Bias is a systematic error that occurs because of incorrect algorithm assumptions. For example, if the algorithm only had information on an apple and no other fruit, it would assume that an apple is the only type of fruit. Because of bias, AI tools like chatbots are more likely to give specific responses over others, even when those answers may be false.
Big Data	An enormous data set too large to process with traditional computing. AI software can analyze these large databases through data mining to identify patterns and draw conclusions. Access to big data allows AI solutions to respond with more intelligence and deliver more human-like interactions.

Burst Test	A method used to evaluate the strength and performance of materials or products under high-pressure conditions. It involves subjecting the material or product to increasing internal pressure until it reaches its bursting point. This test helps determine the maximum pressure the material can withstand before failure, providing valuable insights into its durability, reliability, and safety in real-world applications.
BYOC	Bring Your Own Carrier refers to the practice of allowing organizations to select and integrate their own telecommunications service providers rather than being limited to the options offered by the data center. This enables greater flexibility and customization of services, such as voice, data, and internet connectivity, according to the organization's specific needs and preferences. BYOC can offer benefits like cost savings, improved performance, and the ability to leverage existing relationships with carriers while still utilizing the data center's infrastructure for other services.
CCaaS	Contact Center as a Service is a cloud-based customer experience solution that allows companies to utilize contact center capabilities without the need to own, host, or maintain the underlying infrastructure. CCaaS providers offer a range of functionalities, including but not limited to call routing, customer interaction analytics, workforce optimization, and omnichannel support. This model offers scalability, flexibility, and cost-efficiency, enabling organizations to focus on improving customer service while reducing capital expenditure and operational costs.
CCP	Concurrent Call Path refers to the number of simultaneous voice conversations that a telecommunication system, often within a contact center, can handle at any given time. This metric is crucial for determining the capacity and scalability of a phone system, dictating how many callers can be accommodated without experiencing busy signals or dropped calls. Understanding and planning for the appropriate number of concurrent call paths is vital for ensuring efficient operations, optimal customer experience, and cost-effective utilization of telecommunication resources.

CCPA	The California Consumer Privacy Act is a state-level data protection law that went into effect on January 1, 2020, granting California residents enhanced privacy rights and consumer protection regarding their personal data. The legislation provides individuals with the right to know what data is being collected about them, the right to delete personal data held by businesses, and the right to opt out of the sale of their data. Companies that do business in California and meet certain criteria are required to comply with the CCPA, and non-compliance can result in substantial fines and legal penalties.
CES	Customer Effort Score is a single-item metric that measures how much effort a customer has to exert to resolve an issue, a request, or a question.
Chatbot	An AI program designed to simulate human conversation and provide automated responses to users. It utilizes natural language processing techniques to understand and interpret user input, allowing it to engage in interactive conversations. Chatbots are employed in various applications, such as customer support, virtual assistants, and information retrieval, offering a convenient and efficient way to interact with computer systems through conversation.
ChatGPT	An advanced conversational AI model developed by OpenAI, based on the GPT (Generative Pre-trained Transformer) architecture. It leverages deep learning to understand and generate human-like text responses across a wide range of topics and contexts. With its extensive training on diverse internet text, ChatGPT can engage in meaningful conversations, offer suggestions, and provide information on various subjects.
Classification	A fundamental task in machine learning that involves assigning a category or label to a given input data point based on its characteristics. In other words, classification is a supervised learning method that allows machines to learn how to classify new instances into a pre-defined set of categories based on the features present in the data.

CML	Continuous Machine Learning helps with continuous improvements. CML's most basic application is in circumstances where the data distributions remain constant, but the data is continuous. It automates your Machine Learning process, such as model training and evaluation, comparing trials throughout your project history, and monitoring dataset changes. If you are familiar with Netflix's recommender system, which has an "Up Next" feature that plays shows similar to the ones you've recently watched, then you have seen a CML model in action.
CMS	The Call Management System is a software solution designed to collect and analyze real-time and historical data related to contact center operations. It enables organizations to monitor key performance indicators (KPIs), such as call volumes, handling times, and service level metrics, providing insights that help optimize agent productivity and customer experience. Through its suite of reporting and analytical tools, CMS assists in resource planning, performance evaluation, and strategic decision-making in the contact center environment.
CNN	Convolutional Neural Network is a class of deep neural networks primarily used in the processing and analysis of visual data, such as images and videos. The architecture is designed to automatically and adaptively learn spatial hierarchies of features through its multiple layers, which include convolutional layers, pooling layers, and fully connected layers. CNNs have become the de facto standard for tasks like image recognition, object detection, and various other computer vision applications due to their ability to handle high-dimensional input data and produce accurate results efficiently.

Conversational AI	The use of AI technologies to enable natural language interactions between humans and machines. It involves the development of intelligent systems capable of understanding and generating human-like speech or text. Conversational AI encompasses various technologies, such as chatbots, virtual assistants, and voice recognition systems, aiming to provide seamless and interactive communication experiences for users across different platforms and applications.
CPU	Central Processing Unit, called the "central" or "main" processor, is a complex set of electronic circuitries running the machine's operating system and apps. It is the primary component of a computer that acts as its "control center."
CRM	Customer Relationship Management system is a technology platform that centralizes, automates, and systematizes interactions between a business and its customers. It serves as a repository for customer information, purchase history, service interactions, and other relevant data, thereby enabling organizations to manage relationships, identify opportunities, and optimize customer experiences.
CSAT	Customer Satisfaction is a commonly used metric indicating customer satisfaction with a company's products or services. It's measured through customer feedback and expressed as a percentage (100% would be excellent – 0% would be abysmal).
CTP	Customer Technology Platform is a comprehensive suite of digital tools and systems designed to enhance the interaction between a business and its customers. It integrates various functionalities such as customer relationship management (CRM), communication channels (e.g., email, chat, and social media), and analytics to provide a seamless and personalized customer experience.
Data Drift	The phenomenon where the statistical properties of the target variable, which the machine learning model is trying to predict, change over time in unforeseen ways, causing the model's performance to degrade. This occurs when the data used for training no longer represents the current environment where the model is deployed, leading to inaccuracies and predictive errors.

Data Labeling	Data labeling refers to the process of annotating raw data, such as images, text, or audio, with informative tags or labels to create a labeled dataset. These annotations serve as ground truth, enabling supervised learning algorithms to train on this data to recognize patterns, make predictions, or classify new, unlabeled data. The quality and accuracy of the data labeling process are crucial for the performance and reliability of the AI models trained on such datasets.
DBMS	Database Management Systems are designed to store, manage, and manipulate databases, which can include various types of data such as text, numbers, multimedia, and more. The DBMS provides an interface between the database and the users or the application programs, ensuring that the data is consistently organized and remains easily accessible. Functionality generally includes data retrieval, insertion, update, and deletion, as well as various administrative operations like backup, security, and data integrity.
Deep Learning	A specialized subset of machine learning that mimics the neural networks of the human brain to process data and create patterns for decision-making. It utilizes multiple layers of artificial neural networks to automatically learn to perform tasks without human intervention, essentially training itself by analyzing various forms of data. Deep learning techniques are particularly effective for complex tasks like image and speech recognition, natural language processing, and autonomous driving, often outperforming traditional machine learning algorithms in these domains.
Dialog Management	The process of orchestrating and controlling the flow of conversation between a computer system and a user. It involves coordinating various components, such as natural language understanding, dialog policies, and system responses to enable effective communication. Dialog management systems utilize state tracking, user intent recognition, and response generation to maintain context, handle user requests, and provide appropriate and coherent system responses throughout the dialogue.

Dialogue Logic	The systematic rules and principles that govern the flow of conversation and interaction between participants in a dialogue. It encompasses the logical structure and coherence of the dialogue, including the organization of topics, the sequencing of turns, and the rules for exchanging information and responses. Dialogue logic aims to ensure meaningful and effective communication by establishing guidelines for logical reasoning, turn-taking, and the coherent exchange of ideas.
DID	Direct Inward Dialing refers to a service allowing a company to allocate individual phone numbers to specific lines, extensions, or employees without requiring separate physical phone lines. These numbers are mapped to existing phone lines via a Private Branch Exchange (PBX) system, enabling direct, targeted routing of incoming calls. DID facilitates efficient call management and personalization, allowing for streamlined internal and external communications without the need for a centralized reception or switchboard.
DL	Deep Learning is a subset of machine learning that utilizes artificial neural networks with multiple layers to learn and extract high-level representations from complex data. Deep learning models can recognize intricate patterns in pictures, text, sounds, and other data to produce accurate insights and predictions. Deep learning methods automate tasks that typically require human intelligence, such as transcribing a sound file into text.
Dynamic AI	The use of artificial intelligence systems that can adapt and evolve in real-time based on changing environments or data inputs. It involves the ability of AI models to update their internal representations, decision-making processes, or behavior dynamically. Dynamic AI enables systems to respond effectively to new information, handle varying conditions, and improve their performance over time, making them more flexible, robust, and capable of learning from dynamic and evolving situations.

Dynamic AL	Dynamic Active Learning is an approach used in machine learning and data annotation where the selection of informative samples for labeling is adapted dynamically during the learning process. It involves iteratively updating the sample selection criteria based on the current state of the model and the available labeled data. Dynamic AL aims to maximize the learning efficiency by prioritizing the labeling of samples that are expected to provide the most significant improvement to the model's performance, leading to faster convergence and reduced annotation efforts.
EQ	Emotional Quotients, or Emotional Intelligence (EI), acquire data through real-world data, speech science, and deep learning algorithms. The data is processed and compared to other data points to detect important emotions like fear and joy. After finding the correct emotion, the computer interprets it and what it might mean in each situation. As the emotion database expands, Emotional Intelligence becomes more adept at recognizing the subtleties of human communication.
ETL	Extract, Load, and Transform is a data integration process that begins by extracting data from various sources and then directly loading this raw data into a data storage system, typically a data warehouse or data lake. Once stored, the data undergoes the transformation process, where it is cleansed, enriched, and restructured to meet the business's analytical or operational needs.
F1 Score	A metric used to evaluate the performance of binary classification models, offering a balance between precision and recall. It is calculated as the harmonic mean of the precision and recall, where precision is the number of true positive results divided by the number of all positive results, and recall is the number of true positive results divided by the number of positive results that should have been returned. The F1 Score ranges from 0 to 1, where a score of 1 indicates perfect precision and recall, and a score of 0 indicates the worst possible performance.

FAQ	Frequently Asked Questions generally provide information on frequent questions or concerns. They are often organized in articles, websites, email lists, and online forums where common questions tend to recur—for instance, posts or queries by new users related to common knowledge gaps.
FCR	First Contact Resolution is a metric used to measure customer inquiries or problems resolved on the first call or contact with a representative or agent. FCR is one of the most commonly measured metrics in the contact center industry. Ideally, the FCR definition means no repeat calls or contacts are required from the initial call or contact reason from a customer perspective.
Feedback Data	In predictive machines, feedback data refers to the information or evaluations provided to the model after making predictions or decisions. It is used to assess the performance and accuracy of the model's outputs. Feedback data can include ground truth labels, user ratings, error metrics, or other forms of feedback that help refine and improve the model's predictions through iterative learning processes.
FL	Federated Learning is a machine learning approach that allows a model to be trained across multiple decentralized devices or servers holding local data samples without exchanging them. Instead of sending raw data to a centralized server, local models are trained on each device or server. The model updates are then sent to a central server, aggregating them to produce a global model. This approach aims to improve privacy, security, and efficiency, as sensitive data remains on the local device and only relevant model information is communicated.

Form-Based Model	A type of AI model designed to interact with users through pre-defined form-based inputs and responses. It typically involves structured input formats such as fillable forms or questionnaires, where users provide specific information or answer predefined questions. AI form-based models utilize natural language processing and pattern-matching techniques to understand and process the user's input, generating appropriate responses or actions based on the form data provided. These models are commonly used in applications like surveys, data collection, and automated customer support.
Fourth Industrial Era, The	Also known as the fourth revolution in industry because it brings about a significant change in how industries function, much like the previous industrial revolutions. It uses sophisticated algorithms and computing capabilities to automate processes, enhance decision-making, and improve efficiency and productivity in various sectors. Like the earlier industrial revolutions, AI's influence on industries is predicted to be profound and disruptive, transforming society's landscape.
FSM	Finite State Machines, also known as Finite State AL models, are computational models that consist of a finite number of states and transitions between those states. Each state represents a specific condition or configuration of the system, and the transitions represent the actions or events that cause the system to move from one state to another. Finite State AI models are often used in decision-making processes or simple rule-based systems, where the current state and the input received determine the behavior of the AI.
GAI	Generative AI is a class of AI techniques that focuses on generating new content or data rather than just analyzing or classifying existing information. It involves training models to learn patterns and structures from existing data and to use that knowledge to generate new, original content. Generative AI has applications in various domains, such as image synthesis, text generation, music composition, and even video generation, enabling the creation of realistic and novel outputs.

GAN	Generative Adversarial Networks are a type of neural network architecture consisting of two components, a generator and a discriminator, that compete against each other to generate realistic data samples.
GDPR	General Data Protection Regulation is a comprehensive data protection law enacted by the European Union in 2018, replacing the Data Protection Directive of 1995. It aims to standardize data protection laws across EU member states and to protect the privacy and personal data of EU citizens, including the control and processing of such data. Organizations that handle personal data, irrespective of their geographical location, are obligated to comply with GDPR provisions, with penalties for non-compliance that can be as high as 4% of the company's annual global revenue or €20 million, whichever is greater.
GPT	Generative Pre-trained Transformers are a family of neural network models that uses the transformer architecture and is a crucial advancement in generative AI applications such as ChatGPT, with the ability to create human-like text and content and conversationally answer questions.
GPU	Graphics Processing Unit is a hardware component that accelerates machine learning models' training and inference processes. GPUs excel at performing parallel computations, making them highly efficient for training deep neural networks and processing large datasets. With their massive parallelism and high memory bandwidth, GPUs have become a crucial tool in AI, enabling faster model training and improved performance in various AI applications.
Hallucination	The phenomenon where artificial intelligence systems, particularly generative models, produce imaginative or unrealistic outputs, diverging from the intended or expected results. It can occur when AI models generate highly creative content that lacks fidelity to the real world. AI hallucinations can be observed in various applications, such as image synthesis or text generation, where the models generate outputs that exhibit imaginative elements or distortions not present in the training data.

HIPAA	Health Insurance Portability and Accountability Act of 1996 (HIPAA or the Kennedy–Kassebaum Act) is a United States Act of Congress enacted by the 104th United States Congress and signed into law by President Bill Clinton on August 21, 1996. It modernized the flow of healthcare information and stipulated how personally identifiable information (PII) maintained by the healthcare and healthcare insurance industries should be protected from fraud and theft.
Hyperparameter	A configuration variable that is external to the model and whose value is set prior to the commencement of the training process. Unlike model parameters, which are learned directly from the training data, hyperparameters are not learned from the data but are set a priori to guide the learning process. Common examples of hyperparameters include the learning rate in gradient boosting, the regularization term in logistic regression, and the number of hidden layers in a neural network.
IHT	Interaction Handling Time refers to the duration it takes to handle a customer interaction or query from start to finish in a contact center or customer service environment. It includes the time spent on activities like greeting the customer, gathering information, providing assistance, resolving issues, and concluding the interaction. Monitoring and minimizing IHT is crucial for improving operational efficiency and customer satisfaction in contact center operations.
Inference	The process of using a trained model to make predictions or draw conclusions from new, unseen data. It involves applying the learned patterns and relationships from the training phase to make informed decisions on input data. During inference, the model takes in input data, processes it, and produces the desired output, such as classification labels or regression values, based on its learned knowledge.

Input Data	In predictive machines, input data refers to the information or features provided to the machine learning model for making predictions or decisions. It is the data that the model uses as input to learn patterns and relationships. The input data can vary depending on the specific problem. Still, it typically includes relevant attributes, variables, or measurements that are expected to influence the outcome or prediction made by the model.
Intent	The goal a human has when interacting with a machine. For instance, when a customer asks a chatbot about the location of their package, an AI tool would recognize the user's intent as requesting information about their order status. Identifying a user's intent enables a chatbot to generate specific responses tailored to a person's unique needs.
IVA	Intelligent Virtual Assistant is a software program or application that uses artificial intelligence and natural language processing to interact with users and provide them with assistance or information. IVAs are designed to simulate human-like conversations and can understand and respond to user queries or commands. They are commonly used in customer service, virtual agents, and chatbot applications to provide automated support and enhance user experiences.
IVR	Interactive Voice Response is an automated telephony system that interacts with callers through voice or keypad inputs. It uses pre-recorded voice prompts and menu options to guide callers through various options and route them to the appropriate department or information. IVRs are commonly used in customer support, call centers, and phone-based services to handle a high volume of calls efficiently and provide self-service options to callers.
JSON	JavaScript Object Notation is a lightweight data-interchange format that is easy for humans to read and write and easy for machines to parse and generate. It is based on a subset of the JavaScript language. It is often used for transmitting structured data over a network, commonly in web applications, to exchange data between a client and a server.

KB	Knowledge Base is a set of data available for a program to access to perform a task or give a response. The larger the knowledge base an AI application has access to, the more comprehensive the range of problems it can solve. AI programs can only pull from the knowledge base it has access to.
K-Fold Cross-Validation	A statistical technique used in machine learning to assess the performance of a predictive model. In this method, the original dataset is randomly partitioned into 'k' equally-sized or nearly equal subsamples; of these, a single subsample is retained as the validation set, and the remaining 'k-1' subsamples are used as the training set. The cross-validation process is then repeated 'k' times, each time with a different subsample serving as the validation set, and the model's performance is averaged over all 'k' runs to provide a more robust evaluation.
KPI	Key Performance Indicators are metrics used to assess the performance and effectiveness of a contact center in delivering customer service and achieving business goals. Common KPIs include average handling time (AHT), which measures the average duration of customer interactions; first call resolution (FCR), which tracks the percentage of customer issues resolved in a single contact; and customer satisfaction (CSAT) scores, which gauge customer satisfaction with the service received. Monitoring and improving these KPIs can help organizations enhance operational efficiency, customer experience, and overall contact center performance.
LLM	Large Language Models are deep-learning algorithms that recognize and generate content after training on massive amounts of data. The larger the dataset is, the more effective a language model will be at understanding, translating, and predicting text. LLMs utilize deep learning techniques like transformer architectures to generate human-like text and understand natural language. LLMs can perform various language-related tasks, including text generation, translation, summarization, and question answering, and have been influential in advancing natural language processing applications.

LLMOps	Large Language Models (LLMs) Operations (Ops) in Artificial Intelligence (AI) involve the systematic management and maintenance of LLM systems to ensure they function optimally within their application environment. This includes tasks such as monitoring model performance, managing resource allocation, and implementing updates or improvements based on evolving data inputs and user feedback. LLMOps also encompasses the ethical and security considerations necessary to maintain the integrity and reliability of these models, ensuring they adhere to established guidelines and regulations.
Load Test	A type of performance testing that evaluates the behavior and performance of a system under specific anticipated loads or stress conditions. It involves subjecting the system to simulated user activity or traffic to measure its response time, throughput, and scalability. Load tests help identify potential bottlenecks, performance limitations, or areas of improvement in the system, ensuring it can handle expected loads effectively and maintain optimal performance.
LOB	Line Of Business refers to a distinct set of activities, products, or services that form a recognizable segment of a company's overall operations, often catering to a specific market or type of customer. It represents a focused area of business that has its own strategy, revenue generation models, and management structures, distinguishing it from other business lines within the same organization. The concept is critical for strategic planning and resource allocation, as companies often assess the performance and growth potential of each LOB to optimize profitability and competitive positioning.
LSTM	Long Short-Term Memory is a type of recurrent neural network (RNN) architecture specifically designed to address the vanishing gradient problem and capture long-term dependencies in sequential data. It utilizes memory cells, input, and forget gates to regulate the flow of information, making it capable of learning and retaining information over long sequences. LSTMs have proven effective in various tasks involving sequential data, such as speech recognition, machine translation, and sentiment analysis.

ML	Machine Learning is a branch of artificial intelligence that focuses on developing algorithms and models that allow computers to learn from and make predictions or decisions based on data. It involves using statistical techniques and algorithms to automatically enable computers to identify patterns and extract meaningful insights from large datasets. Machine learning algorithms can adapt and optimize their predictions or actions over time by iteratively improving their performance through data exposure.
Model	The model defines the relationship between the input data, which we call features, and what the model is trying to predict, called the labeled data. The data that is labeled is evidence. Labeled data could be the square footage of a house or the total sales in a day. It's the evidence that gets collected and submitted. The models then train with these datasets; the more data you give a model, the better it predicts over time.
NCP	Non-Player Characters are characters in video games or virtual environments controlled by the game's artificial intelligence rather than a human player. They are designed to interact with the player or other characters, often serving specific roles within the game's storyline, quests, or gameplay mechanics. NPCs can range from friendly allies to neutral bystanders or hostile enemies, providing a dynamic and immersive experience for the player.
NER	Named Entity Relationships is an NLP method that extracts information from text, detecting and categorizing pertinent information known as "named entities." Named entities refer to the key subjects of a piece of text, such as names, locations, companies, events, and products, as well as themes, topics, times, monetary values, and percentages.

Neural Networks	A class of machine learning models inspired by the structure and function of the human brain. They consist of interconnected nodes, called neurons, organized into layers. These networks learn from data by adjusting the weights and biases of the connections between neurons, enabling them to capture complex patterns and relationships in the data and make predictions or decisions. Neural networks have shown remarkable performance in various domains, including image recognition, natural language processing, and pattern classification.
NLG	Natural Language Generation is a field of AI that focuses on generating human-like natural language text or speech based on structured data or other input forms. It is the opposite of NLU transforming structured information into coherent and understandable human spoken words mimicking natural language patterns and conventions. NLU can be combined with other AI technologies, such as natural language understanding and dialog systems, to create interactive conversational agents or chatbots.
NLP	Natural language processing has existed for over 50 years and has roots in linguistics. In AI, NLP is a program's ability to interpret written and spoken human language, allowing computers to understand text and spoken words like human beings, including their tone and intent. NLP combines computational linguistics for rule-based modeling of human language with statistical, machine learning, and deep learning models. There are two main phases to natural language processing: data preprocessing involves preparing and "cleaning" text data for machines to be able to analyze it, and algorithm development. NLP enables chatbots to detect customer sentiment, including determining if the customer is frustrated, complaining, or simply completing a request.

NLU	Natural Language Understanding is a branch of AI focusing on the interaction between computers and human language. It is a subfield of NLP that enables computers to understand, interpret, and derive meaning from language similarly to humans. NLU involves developing algorithms, models, and systems to process and comprehend natural language inputs, such as text or speech, extract relevant information, understand the context, and derive the intended meaning.
NPS	Net Promoter Score is a metric used in contact centers to measure customer loyalty and satisfaction. It is based on the question, "On a scale of 0 to 10, how likely are you to recommend our company/service to a friend or colleague?" Customers are then categorized into promoters (rating 9-10), passives (rating 7-8), or detractors (rating 0-6). The NPS is calculated by subtracting the percentage of detractors from the percentage of promoters, providing an overall indication of customer sentiment and loyalty.
OpenAI	A research organization focused on artificial intelligence (AI) and machine learning. It aims to develop AI technologies that benefit humanity while ensuring safety and ethical considerations. OpenAI has produced models, like GPT and DALL-E, that have advanced natural language processing and image generation.
PBX	Private Branch eXchange is a small version of a telephone company's central office. PBXs can handle both inbound and outbound calls but are more flexible and can be programmed to meet your business requirements.
PoC	Proof of Concept is an experimental demonstration designed to ascertain the feasibility and practicality of a proposed idea, project, or system. It serves as an initial stage where key technical and operational aspects are tested, but without the requirement to be fully functional or comprehensive. By validating the core mechanisms and potential benefits, a PoC helps decision-makers assess whether the idea is worth investing in for full-scale development.

Prediction Machines	Computational systems or algorithms designed to make predictions or forecasts based on available data. They utilize machine learning and statistical techniques to analyze patterns, correlations, and trends in data and generate predictions about future outcomes. Prediction machines have diverse applications across industries, including finance, healthcare, weather forecasting, and sales forecasting, providing valuable insights for decision-making and planning.
Predictive Analytics	A machine learning technique to predict future outcomes or behaviors using historical data and statistical algorithms. It involves analyzing past patterns and trends to make informed predictions about future events, behaviors, or probabilities. By leveraging data-driven insights, predictive analytics enables organizations to anticipate outcomes, make proactive decisions, and optimize strategies across various business, finance, marketing, and healthcare domains.
Primary Data	Original, unmediated data collected directly from the environment, subjects, or processes under study, specifically for the purpose of training, validating, or testing the algorithm. Primary data is raw and closely represents the attributes or variables that the algorithm aims to analyze or predict. The quality, reliability, and appropriateness of the primary data source are critical factors in determining the accuracy and efficacy of the resulting AI model. See Secondary Data.
Probabilistic Model	A mathematical framework used to represent uncertainty and randomness in data. It assigns probabilities to different outcomes or events based on available information or prior knowledge. By capturing the probabilistic relationships between variables, a probabilistic model enables reasoning, prediction, and inference about uncertain quantities and allows for probabilistic reasoning in decision-making processes.

RAG	Retrieval-Augmented Generation represents an advanced paradigm that synergizes the capabilities of retrieval-based and generative models to enhance the generation of text. By integrating a retrieval mechanism, RAG models can access a vast external knowledge base, allowing them to pull in relevant information during the generation process, thereby improving the accuracy and relevance of the output. RAG not only enhances the model's ability to produce contextually informed responses but also broadens its applicability across diverse domains requiring nuanced and detailed content generation.
RBAC	Role-Based Access Control is a security framework in which permissions and access to resources are granted based on predefined roles within the organization. Under this model, each role is associated with a specific set of permissions and access rights, which are then inherited by users assigned to that role. RBAC simplifies the management of access controls by allowing administrators to manage roles rather than individual users, thus making it easier to enforce security policies and manage personnel changes within the contact center.
Redaction	The process of removing or obscuring sensitive or confidential information from a document, audio file, or other data source before it is published, shared, or archived. The primary objective is to protect individual privacy, comply with regulations, or meet security requirements while still making the non-sensitive portions of the information available for use. In the context of contact centers, redaction is often employed to ensure that customer data and interactions are compliant with privacy laws, such as the GDPR or HIPAA. See Anonymization.

Regression Testing	A software testing technique to ensure that recent changes or modifications to a system have not introduced new defects or caused unintended side effects in previously functioning features. It involves retesting the existing test cases to verify the system's behavior and confirm that it still operates as expected after the changes. Regression testing helps maintain the stability and reliability of the software by catching any regressions or issues that might have been introduced during the development process.
Reinforcement Learning	A machine learning paradigm that involves an agent learning to make sequential decisions through interaction with an environment. The agent receives feedback through rewards or punishments based on its actions, allowing it to learn optimal strategies through trial and error. Maximizing cumulative rewards over time, the agent aims to find the best course of action in a given environment, making reinforcement learning well-suited for tasks such as game-playing, robotics, and autonomous decision-making.
Request Technologies	Request Technologies combined with NLG responses allow customers to interact in two-way conversations between the customer and an AI-powered prediction machine.
RLHF	Reinforcement Learning from Human Feedback is an approach in ML where an agent learns from feedback provided by human trainers to improve decision-making capabilities. It involves a human providing evaluations or demonstrations to guide the agent's learning process. RLHF leverages the expertise of human trainers to accelerate the learning process and enhance the performance of the agent in complex and dynamic environments.
RNN	Recurrent Neural Networks are a class of artificial neural networks designed to process sequential data by utilizing feedback connections. They can retain information from previous inputs in their hidden states, allowing them to capture temporal dependencies and contextual information. RNNs have found applications in various tasks such as natural language processing, speech recognition, and time series analysis, where the order and sequence of the data play a crucial role in the analysis and prediction.

ROI	Return on Investment is a financial metric used to evaluate the profitability and effectiveness of an investment, expressed as a percentage. It is calculated by dividing the net gain obtained from the investment by its initial cost and then multiplying the result by 100. ROI serves as a key indicator for businesses and individuals to assess the value generated by an investment relative to its cost, aiding in decision-making processes for future investments or resource allocations. See TCO.
Rule-based Systems	Rule-based Systems are a type of knowledge-based system that utilizes a set of predefined rules to make decisions or perform actions. These systems consist of a rule base containing a collection of if-then statements or conditions and an inference engine that applies these rules to process input data and generate output. Rule-based Systems are commonly used in domains where expert knowledge can be explicitly represented, such as in expert systems, decision support systems, and diagnostic systems.
Secondary Data	A dataset not originally collected for the specific purpose of training, validating, or testing the algorithm in question but repurposed for these functions. Secondary data may have been gathered for other research objectives, administrative records, or prior analyses and is often aggregated, processed, or summarized. While secondary data sources offer the advantages of time and cost savings, their utility in AI applications may be limited by issues related to data relevance, consistency, and completeness. See Primary Data.
Semi-structured Data	Data that falls between structured and unstructured data and contains elements of both. It is characterized by its lack of a rigid structure but does have tags, hierarchies, or other markers that help categorize and organize the information. Formats like XML, JSON, and YAML files are common examples of semi-structured data, where the data is not in a tabular form but contains tags or keys to provide some level of structure and context. See Structured Data, Unstructured Data.

Semi-supervised Learning	A machine learning paradigm that combines supervised and unsupervised learning elements. It involves training a model using a small amount of labeled data and a larger amount of unlabeled data. The labeled data helps the model learn from explicit examples, while the unlabeled data aids in discovering underlying patterns and structures in the data. This approach is useful when labeled data is scarce or expensive to obtain, allowing for more efficient and cost-effective training of models.
SIP	Session Initiation Protocol is a telecommunications protocol widely used for initiating, maintaining, and terminating real-time sessions that involve voice, video, messaging, and other communications applications and services. It functions at the application layer of the Internet Protocol (IP) suite and plays a pivotal role in the control of multimedia communication sessions in Voice over IP (VoIP) networks. By establishing, modifying, or terminating multimedia sessions, SIP serves as a fundamental protocol in the architecture of various forms of internet telephony and converged IP-based communication systems.
SLA	Service Level Agreement is a contractual document that delineates the standards of service a client can expect from a service provider, specifying metrics like response times, resolution times, and availability rates. These agreements serve as a formal understanding between both parties, delineating the scope and quality of the services to be provided. SLAs are critical for establishing mutual accountability and setting clear performance benchmarks, often incorporating penalties for non-compliance and outlining remediation procedures.
Slots	Slots are variables or placeholders used to store information during the processing of input data. Slots are associated with specific states in the model and can be filled with different values as the AI system encounters and processes various inputs. By utilizing slots, the AI model can retain context and make decisions based on the accumulated information within the states.

SLU	Speech Language Understanding is a subfield of natural language processing (NLP) that focuses on interpreting and understanding spoken language by machines. It involves analyzing and processing spoken utterances to extract meaning, intent, and context. SLU techniques are used in applications such as voice assistants, speech recognition systems, and automated call centers to enable machines to comprehend and respond to human speech effectively.
SMPC	Secure Multi-Party Computation is a cryptographic technique that allows multiple parties to collaboratively compute a function over their respective inputs while keeping those inputs private. In essence, SMPC enables computations to occur without requiring any party to reveal their confidential data to others, thus ensuring data privacy and security. This methodology is particularly useful in scenarios where sensitive data must be analyzed collectively without exposure, such as in privacy-preserving medical research, financial services, or secure voting systems.
SMS	Short Message Service is a communication protocol for sending short text messages between mobile devices. It enables users to exchange text-based messages, typically limited to 160 characters per message, over cellular networks. SMS is widely used for personal and business communications, providing a quick and convenient way to send messages between mobile devices.
SQL	Structured Query Language is a domain-specific language used for managing and manipulating relational databases. SQL enables tasks such as querying data, updating data, inserting data, and deleting data from a database, as well as creating and modifying schemas that describe the structure of the database. It serves as the standard interface for various DBMS, including proprietary and open-source solutions.

Static AI	Artificial intelligence systems or models trained on fixed, pre-existing datasets that do not dynamically adapt or update based on real-time information or feedback. These AI models operate based on a predetermined set of rules and patterns and do not actively learn or evolve from new data. Static AI is commonly used in scenarios where the underlying data is relatively stable, and there is no need for continuous learning or adaptation.
Stress Test	A type of performance testing that evaluates the behavior and stability of a system under extreme or peak load conditions. It aims to identify a system's breaking point or maximum capacity and assess its response to heavy traffic or stress factors. By simulating high user volumes or intensive workloads, stress tests help uncover performance bottlenecks, resource limitations, or system failures, providing insights into the system's robustness and resilience.
Structured Data	Data organized into a specific format or schema, such as tables, rows, and columns, which makes it easily searchable and analyzable by computer algorithms. Common examples include relational databases, where data is stored in tables or CSV files where values are separated by commas. The explicit structure and organization of this type of data facilitate efficient querying and analysis, making it highly suitable for traditional data management systems and analytical tools. See Unstructured Data, Semi-structured Data.
Supervised Learning	A machine learning approach where a model is trained using labeled data, consisting of input samples and corresponding output labels. The goal is to enable the model to learn the mapping between input features and their corresponding target values. During training, the model generalizes from the labeled data and can predict unseen data by inferring patterns and relationships learned from the training examples. See Unstructured Learning.

TCO	Total Cost of Ownership is a financial metric used to assess the comprehensive costs associated with acquiring, operating, and maintaining a product or system over its entire lifecycle. TCO encompasses not only the initial purchase price but also ongoing operational costs, maintenance expenses, and any costs related to downtime or inefficiency. By providing a holistic view of expenditure, TCO helps organizations make informed decisions about investments, comparing both short-term and long-term implications. See ROI.
TFN	Toll-Free Numbers are telephone numbers that allow callers to reach businesses or organizations without being charged for the call. The cost of the call is instead borne by the receiving party, usually the business that owns the toll-free number. Businesses, including contact centers, often use these numbers to encourage customer engagement and provide a cost-free channel for customer support, inquiries, or sales.
TPS	Transactions Per Second in an AI interaction refers to measuring how many individual transactions or interactions can be processed by an AI system within one second. It quantifies the system's capacity to handle incoming requests, such as user queries, predictions, or data processing tasks. Higher TPS values indicate a greater ability to handle a larger volume of interactions efficiently, enabling real-time or high-speed AI applications.
TPU	Tensor Processing Units are specialized hardware accelerators designed by Google for machine learning workloads. They excel in processing and manipulating large-scale tensor operations, prevalent in deep learning models. TPUs offer high performance and energy efficiency, enabling faster training and inference times for AI tasks, and they are particularly effective in handling computationally intensive tasks such as neural network training and inference.

Training Data	In predictive machines, training data refers to the labeled dataset used to train a machine learning model. It consists of input samples and their corresponding known output or target values. The model uses Training Data to learn patterns, correlations, and statistical relationships between the input features and the target variable, allowing the model to make accurate predictions on new, unseen data.
Transformative Technology	Innovations that significantly alter the existing market landscape and create new opportunities, often by displacing established technologies or industries. They typically introduce novel approaches, products, or services that offer superior performance, efficiency, or cost-effectiveness compared to traditional solutions. Transformative technologies have the potential to reshape industries, drive market shifts, and bring about transformative changes in business models and consumer behavior.
Transformers	A type of deep learning model architecture that revolutionized natural language processing tasks. They utilize self-attention mechanisms to capture relationships between words or tokens in a sequence, allowing for parallel processing and capturing long-range dependencies. By leveraging attention mechanisms, transformers excel in tasks like machine translation, text generation, sentiment analysis, and more, surpassing the previous sequential approaches and achieving state-of-the-art performance in many language-related tasks.
TSAPI	Telephony Services Application Programming Interface is a set of programming protocols that allows for the integration of computer and telephone systems, commonly employed in Computer Telephony Integration (CTI) environments. Developed by Avaya and Novell, TSAPI enables applications to control and monitor telephony events, facilitating functions such as call routing, queuing, and interactive voice response (IVR). By providing a standardized interface for telephony applications, TSAPI enhances interoperability between different systems and enables more streamlined and efficient communication workflows.

TTS	Text-To-Speech is a technology that converts written text into spoken audio. It involves the synthesis of human-like speech from text input using computational algorithms and linguistic models. TTS systems are utilized in various applications, such as voice assistants, accessibility tools, e-learning platforms, and entertainment, to provide natural and intelligible speech output from written content.
Turing Test	Designed by Alan Turing in 1950, the Turing Test is a test of a computer's ability to display intelligence that is indistinguishable from human intelligence. The test theorized that the software's intelligent behavior could be measured against human intellectual efficiency. The software is intelligent when a human does not know if they are chatting with the software or with another human. Unfortunately, access to computers combined with their functional limitations blocked the development of any proof of concept until recently.
Unstructured Data	Information that lacks a predefined schema or structure, making it challenging to categorize, query, or analyze through conventional database algorithms. This type of data often includes text, images, videos, and audio files, among other formats, which do not fit neatly into tabular structures like rows and columns. Due to its complexity and lack of organization, specialized tools and techniques, such as natural language processing or machine learning algorithms, are often required to analyze and extract meaningful insights from unstructured data. See Structured Data, Semi-structured Data.
Unsupervised Learning	A machine learning approach where the model learns patterns and structures in data without explicitly labeled examples. It aims to uncover hidden relationships, clusters, or patterns within the input data. Unlike supervised learning, there are no predefined target labels, and the model relies on inherent structures or similarities in the data to identify meaningful insights and make sense of the data. See Structured Learning.

VAR	Variational Autoencoders are generative models in artificial intelligence that combine elements of both autoencoders and probabilistic models. They aim to learn a compact and continuous latent representation of input data by simultaneously training an encoder and decoder network. VAEs introduce a probabilistic approach to autoencoders, allowing for the generation of new data samples by sampling from the learned latent space, making them useful for tasks like data generation, dimensionality reduction, and unsupervised learning.
Voice Bot	An artificial intelligence system that interacts with users through voice commands and responses. Also known as a voice assistant or voice-enabled chatbot. It leverages natural language processing and speech recognition technologies to understand and interpret spoken input. Voice bots are commonly used in applications like virtual assistants, customer support, and smart home devices, offering a hands-free and intuitive way to access information and perform tasks using voice interactions.
Voice-First	Where the voice interaction takes precedence over other forms of input, such as text or touch, and is prioritized as the primary mode of communication between the customer and the system.
VR	Virtual Reality is the technology that creates simulated environments and experiences, often using computer-generated visuals and audio. It aims to immerse users in a virtual world, stimulating their senses and enabling interactive experiences. AI techniques can be employed in VR systems to enhance aspects such as realistic graphics, natural language interactions, and intelligent behavior of virtual entities, making the virtual experience more immersive and engaging.

VUI	Voice User Interface is a technology that allows users to interact with a system through voice or speech commands rather than traditional input methods like touch, keyboard, or mouse. VUIs are increasingly employed in customer service applications in contact centers. The primary objective of a VUI is to provide a more natural, efficient, and accessible means for users to engage with technology, often leveraging artificial intelligence and natural language processing to understand and respond to user requests.
Wake Word	Also known as a trigger word or hot-word, the term refers to a specific word or phrase that acts as a signal to activate a voice-controlled system or virtual assistant. It is a starting point for initiating a conversation or interaction with the system. Request technologies listen for user input in a continuous process. This analyzes audio or text data to detect if the wake word has been spoken or mentioned. Once the wake word is recognized, the system activates and begins capturing and processing subsequent commands or queries. Well-known examples of a wake word are "Hey Siri" in Apple products and "Alexa" used by Amazon.
Waterfall Methodology	A linear and sequential approach to software development, where each phase of the project is completed before moving on to the next. It typically consists of distinct phases, such as requirements gathering, design, implementation, testing, and maintenance, with little room for iteration or flexibility once a phase is completed. While it offers clarity and structure, the waterfall model can lead to difficulties in accommodating changes late in the development process and may result in longer development cycles. See Agile Methodology.

WFM	Workforce Management is the systematic process that involves planning, scheduling, and monitoring the allocation of human resources to ensure optimal efficiency and productivity within an organization. It encompasses a range of activities, including demand forecasting, staff scheduling, time and attendance tracking, and performance evaluation. By integrating these elements, workforce management aims to balance operational requirements with employee capabilities, thereby enhancing organizational performance and employee satisfaction.
XAI	Explainable AI is an approach in artificial intelligence that emphasizes the transparency and interpretability of AI models and their decision-making processes. It aims to provide understandable explanations for the outcomes or predictions generated by AI systems. By enabling humans to comprehend and trust the reasoning behind AI decisions, explainable AI promotes accountability, fairness, and the identification of potential biases or errors in the decision-making process.
XML	Extensible Markup Language is a text-based markup language that enables the structured representation of data. Developed by the World Wide Web Consortium (W3C), XML provides a way to encode documents or data fields, often using tags to delineate and nest information hierarchically. While not specific to any particular software or hardware, XML is highly extensible and serves as a common standard for data interchange between disparate systems, supporting a wide range of applications in fields such as web development, document storage, and data transmission.

References

Section 1. Introduction

Rule 1. Where to Begin

1. *Is your AI center of excellence still a center of experimentation?*. (2023). Deloitte. Retrieved November 20, 2023, from https://www2.deloitte.com/content/dam/Deloitte/us/Documents/consulting/us-is-your-ai-center-of-excellence-still-center-of-experimentation.pdf

2. Goel, N., Roy, S., & Kejriwal, R. (2023, May 11). *10 Levers of Success to Build an Effective Center of Excellence*. Zinnov. https://zinnov.com/centers-of-excellence/10-levers-of-success-for-a-center-of-excellence-blog/

3. Davenport, T. H., & Dasgupta, S. (2019, January 16). *How to Set Up an AI Center of Excellence*. Harvard Business Review. https://hbr.org/2019/01/how-to-set-up-an-ai-center-of-excellence

4. *How to Build an AI Center of Excellence*. (2023, March 3). DDN. https://www.ddn.com/resources/whitepapers/how-to-build-an-ai-center-of-excellence-to-accelerate-ai-expertise/

Rule 2. What Is an AI-CoE?

5. Brashear, J., & Ammanath, B. (2020). *Taking AI to the next level: Harnessing the full potential and value of AI while managing its unique risks*. Deloitte. Retrieved November 20, 2023, from https://www2.deloitte.com/content/dam/

Deloitte/us/Documents/process-and-operations/us-taking-ai-to-the-next-level.pdf

6. *Exploring the uses and benefits of a Center of Excellence.* (n.d.). Lucidchart. https://www.lucidchart.com/blog/exploring-the-uses-and-benefits-of-a-center-of-excellence#:~:text=A%20CoE%20helps%20create%20a,and%20push%20each%20other%20forward

Rule 3. Why an AI-CoE for Contact Centers?

7. Kafka, M. (2022, June 13). *Center of excellence: What is it and why create one?.* OneSpan. https://www.onespan.com/blog/center-excellence-why-create-one

8. *6 Steps to Building a Center of Excellence.* (2022, October 27). Bizagi Site. https://www.bizagi.com/en/blog/digital-transformation-strategy/6-steps-to-building-a-center-of-excellence

9. Vashishta, V. (2023, February 2). *What Is an AI Center of Excellence?.* AtScale. https://www.atscale.com/blog/what-is-an-ai-center-of-excellence/

Rule 4. Define the Value of the AI-CoE

10. Khalaf, R., Srouji, Y., Haddad, S., & Ghanem, S. (2022). *Centers of Excellence: Setting up for success.* Deloitte. Retrieved November 23, 2023, from https://www2.deloitte.com/content/dam/Deloitte/xe/Documents/consulting/me_centers-of-excellence-(COEs).pdf

Rule 5. View Your CX from the Outside In

11. Alvarez, G., Gianni, A., Brand, S., & Lowndes, G, M. (2023). *Hype Cycle for the Future of Enterprise Applications, 2023: Customer Technology Platform.* Gartner, G00790920.

Rule 6. Centralize or Federate Your AI-CoE

12. *Is Your AI Center of Excellence Still a Center of Experimentation?.* (2023). Deloitte. Retrieved January 15, 2024, from https://www2.deloitte.com/content/dam/Deloitte/us/Documents/consulting/us-is-your-ai-center-of-excellence-still-center-of-experimentation.pdf

Rule 7. Mind Map Your AI-CoE

13. Vashishta, V. (2023, February 2). *What Is an AI Center of Excellence?.* AtScale. https://www.atscale.com/blog/what-is-an-ai-center-of-excellence/

Section 2. Align Your Strategic Objectives

14. Easton, S., & Dures, D. (n.d.). *Strategic Alignment: The Ultimate Guide.* Transparent Choice. https://www.transparentchoice.com/strategic-alignment

15. Gangnes, J. T. (2022, June 28). *Strategic Alignment: What It Is and How to Achieve It.* ThoughtExchange. https://thoughtexchange.com/blog/strategic-alignment/

16. Khalaf, R., Srouji, Y., Haddad, S., & Ghanem, S. (2022). *Centers of Excellence: Setting up for success.* Deloitte. Retrieved November 23, 2023, from https://www2.deloitte.com/content/dam/Deloitte/xe/Documents/consulting/me_centers-of-excellence-(COEs).pdf

Rule 8. Align Your Customer Objectives

17. Goel, N., Roy, S., & Kejriwal, R. (2023, May 11). *10 Levers of Success to Build an Effective Center of Excellence.* Zinnov. https://zinnov.com/centers-of-excellence/10-levers-of-success-for-a-center-of-excellence-blog/

18. Contact Center – Center of Excellence. (2020, October 8). *Fifteen Plays of Our Contact Center Approach.* GSA - IT Modernization Centers of Excellence. https://coe.gsa.gov/2020/10/08/cc-update-5.html#Play9

Rule 9. Align Your Business Objectives

19. *Everything You Need to Know About Centers of Excellence.* (2022, August 3). Catalant. https://catalant.com/coe-everything-you-need-to-know-about-centers-of-excellence/

20. Tadwalkar, S. V. (2008). Create centre of excellence (CoE) for better business. *Ubiquity*, *2008*(9), 1–9. https://doi.org/10.1145/1366313.1361366

Rule 10. Align Your Technology Objectives

21. Tadwalkar, S. V. (2008). Create centre of excellence (CoE) for better business. *Ubiquity*, *2008*(9), 1–9. https://doi.org/10.1145/1366313.1361366

Rule 11. Align Your Financial Objectives

22. *Aligning Financial Goals with Business Objectives*. (2024, March 16). FasterCapital. https://fastercapital.com/content/Aligning-Financial-Goals-with-Business-Objectives.html#Determining-Financial-Goals-for-Your-Business

23. Maurya, R. (2023, May 4). *Roles and Functions of Management Accountants*. Fundamentals of Accounting. https://fundamentalsofaccounting.org/roles-and-functions-of-management-accountants/

Rule 12. Identify Your Data Sources

24. Dilmegani, C. (2024, Jan 12). *AI Center of Excellence (AI CoE): What it is & how to build in '24*. AIMultiple. https://research.aimultiple.com/ai-center-of-excellence/

25. Davenport, T. H., & Dasgupta, S. (2019, January 16). *How to Set Up an AI Center of Excellence*. Harvard Business Review. https://hbr.org/2019/01/how-to-set-up-an-ai-center-of-excellence

Section 3. Technology and Innovation

26. Mcmillen, D. (2022, October 27). *4 steps for creating a center of excellence (CoE) in your organization*. Enable Architect. https://www.redhat.com/architect/center-of-excellence-coe

27. *CoE: What is a center of excellence?*. (2023, September 14). User Testing. https://www.usertesting.com/blog/coe-center-of-excellence

28. Raval, V. (2023, February 25). *Different Types of "CoEs - Center of Excellence."* LinkedIn. https://www.linkedin.com/pulse/different-types-coes-center-excellence-vision-raval/

Rule 13. Architect Your Cloud

29. *Set Up Your Organization for Cloud Adoption Success.* (n.d.). Gartner. https://www.gartner.com/en/conferences/hub/cloud-conferences/insights/how-to-build-a-cloud-center-of-excellence

30. Mcmillen, D. (2022, October 27). *4 steps for creating a center of excellence (CoE) in your organization.* Enable Architect. https://www.redhat.com/architect/center-of-excellence-coe

Rule 14. Foster Innovation

Rule 15. Proof of Concept

31. *AI Guide for Government.* (n.d.). IT Modernization Centers of Excellence. https://coe.gsa.gov/coe/ai-guide-for-government/print-all/index.html

32. Lach, M. (2022, June 21). *AI Proof of Concept: The Benefits of Kickstarting an AI Software Development Project with AI POC.* Nexocode. https://nexocode.com/blog/posts/ai-proof-of-concept-benefits-of-ai-poc/

Rule 16. Natural Language Processing

33. *What is the future of NLP in AI?.* (n.d.). LinkedIn. https://www.linkedin.com/advice/1/what-future-nlp-ai-skills-artificial-intelligence-qe8re

34. *AI Guide for Government.* (n.d.). IT Modernization Centers of Excellence. https://coe.gsa.gov/coe/ai-guide-for-government/print-all/index.html

35. Infused Innovations, Inc. (2023, May 30). *Unleashing the Power of AI: Establishing an AI Center of Excellence.* LinkedIn. https://www.linkedin.com/pulse/unleashing-power-ai-establishing-center-excellence

Rule 17. Dialog Management

36. Vázquez, A., Zorrilla, A. L., Olaso, J. M., & Torres, M. I. (2023). Dialogue Management and Language Generation for a Robust Conversational Virtual Coach: Validation and User Study. *Sensors,* *23*(3), 1423. https://doi.org/10.3390/s23031423

37. Mantha, M. (2019, March 28). *Conversational AI: Design & Build a Contextual AI Assistant.* Medium. https://towardsdatascience. com/conversational-ai-design-build-a-contextual-ai-assistant-61c73780d10

Section 4. Governance

38. Gotts, I. (2023, February 1). *Center of Excellence: GOVERNANCE.* Elements cloud. https://elements.cloud/blog/coe-governance/

39. Weiss, C. (2019, June). *Selecting Center of Excellence Governance Structures.* Anaplan. https://community.anaplan.com/ discussion/46954/selecting-center-of-excellence-governance-structures

Rule 18. Create a Methodology

40. Infused Innovations, Inc. (2023, May 30). *Unleashing the Power of AI: Establishing an AI Center of Excellence.* LinkedIn. https://www.linkedin. com/pulse/unleashing-power-ai-establishing-center-excellence/

Rule 19. Project Lifecycle

41. *Instant Insights: The AI/Machine Learning.* (2021, October 14). Trust Insights. https://www.trustinsights.ai/insights/instant-insights/ instant-insights-the-ai-machine-learning-lifecycle/

42. Nieto-Rodriguez, A., & Viana Vargas, R. (2023, February 2). *How AI Will Transform Project Management.* Harvard Business Review. https://hbr.org/2023/02/how-ai-will-transform-project-management

43. *A Step-by-Step Approach to Running AI and Machine Learning Projects.* (n.d.). Cognilytica. https://www.cognilytica.com/a-step-by-step-approach-to-running-ai-and-machine-learning-projects/

Rule 20. Standardize Your Design

44. Vartak, M. (2022, March 4). *How to Scale AI in Your Organization.* Harvard Business Review. https://hbr.org/2022/03/how-to-scale-ai-in-your-organization

45. *How to Build an AI Center of Excellence.* (n.d.). DDN. https://www. ddn.com/resources/whitepapers/how-to-build-an-ai-center-of-excellence-to-accelerate-ai-expertise/

46. *AI Center of Excellence: Future-proof your business.* (2023, September 9). Truefoundry. https://www.truefoundry.com/blog/ai-center-of-excellence-how-you-can-future-proof-your-business

Rule 21. Manage Change

47. Contact Center – Center of Excellence. (2020, October 8). *Fifteen Plays of Our Contact Center Approach.* GSA - IT Modernization Centers of Excellence. https://coe.gsa.gov/2020/10/08/cc-update-5.html#Play15

Rule 22. Reuse AI Components

48. Zhou, M. (2021, December 28). *How no-code, reusable AI will bridge the AI divide.* InfoWorld. https://www.infoworld.com/article/3644968/how-no-code-reusable-ai-will-bridge-the-ai-divide.html

49. *Mastering AI Centers of Excellence: Key Considerations for Success.* (2023, October 31). Zinnov. https://zinnov.com/centers-of-excellence/mastering-ai-centers-of-excellence-key-considerations-for-success-blog/

50. Kelly, J. I. (2023, August 23). *6 Powerful Examples of AI in the Contact Center.* Invoca. https://www.invoca.com/blog/examples-ai-contact-center

51. RingCentral Team. (2024, January 7). *Contact center AI: 5 examples for smarter service.* RingCentral Blog. https://www.ringcentral.com/us/en/blog/how-contact-centers-use-ai-strategically/

Section 5. Data and Analytics

52. Lin, A. (2022, September 20). *Five Steps to Building an AI Center of Excellence.* Domino. Retrieved December 20, 2023, from https://domino.ai/blog/five-steps-to-building-the-ai-center-of-excellence

53. Vashishta, V. (2023, February 2). *What Is an AI Center of Excellence?.* AtScale. https://www.atscale.com/blog/what-is-an-ai-center-of-excellence/

Rule 23. Manage Machine Learning

54. *Why Is a Machine Learning CoE Called the Heart of an Intelligent Organization?* (2019, February 14). Zinnov. https://zinnov.com/

centers-of-excellence/machine-learning-centers-of-excellence-a-crucial-growth-lever-for-organizations-blog/

55. Abhirami. (2023, February 28). *Important Role Played by Machine Learning in Contact Centers.* C-ZENTRIX. https://www.c-zentrix.com/blog/role-machine-learning-contact-centers

56. Giovis, R., & Rozsa, E. (2023, July 17). *Transforming customer service: How generative AI is changing the game.* IBM. https://www.ibm.com/blog/transforming-customer-service-how-generative-ai-is-changing-the-game/

Rule 24. Prepare Your Data

57. *Predictive Revenue Intelligence.* (n.d.). ForecastEra. https://forecastera.com/analytics-ai-coe/

58. *Data Preparation for Artificial intelligence (AI).* (n.d.). Clickworker.com. https://www.clickworker.com/customer-blog/data-preparation-for-ai/

59. Foster, E., Walch, K., & Schmelzer, R. (2023, September 7). *How to build a machine learning model in 7 steps.* Enterprise AI. https://www.techtarget.com/searchenterpriseai/feature/How-to-build-a-machine-learning-model-in-7-steps

Rule 25. Model Preparation

60. *AI model governance: What it is and why it's important.* (2023, October 25). Collibra. https://www.collibra.com/us/en/blog/ai-model-governance-what-it-is-and-why-its-important

61. Hietpas, S. (2023, December 19). *A Guide to Preparing Organizational Data for AI.* Core BTS. https://corebts.com/blog/a-guide-to-preparing-organizational-data-for-ai/

Rule 26. Redact Your Data

62. *Data Redaction: What It Is and When to Use It.* (2023, November 28). Informatica. https://www.informatica.com/blogs/data-redaction-what-it-is-and-when-to-use-it.html

63. *Redacting data method (Masking flow).* (2024, January 12). IBM Cloud Pak for Data. https://dataplatform.cloud.ibm.com/docs/content/wsj/governance/dp-redact-method.html?context=cpdaas

64. Devane, H. (2023, April 5). *What Is Data Redaction?*. Immuta. https://www.immuta.com/blog/what-is-data-redaction/

Rule 27. Federate Your Data

65. Ricardo, C. (2003). Database Machines. In H. Bidgoli (Ed.), *Encyclopedia of Information Systems* (pp. 403-410). Elsevier eBooks. https://doi.org/10.1016/b0-12-227240-4/00027-7

66. Laux, M. (2015, October 9). *If you did not already know: "Data Federation."* AnalytiXon. https://analytixon.com/2015/10/09/if-you-did-not-already-know-data-federation/

Rule 28. Large Language Models Operations

67. *How to Use ChatGPT in Your Call Center?*. (n.d.). Nectar Desk. https://www.nectardesk.com/how-to-use-chatgpt-in-your-call-center/

68. *LLMOps CoE: The next frontier in the MLOps Landscape.* (2023, July 20). TrueFoundry. https://blog.truefoundry.com/the-future-is-here-llmops/

Rule 29. It's About Analytics

69. Clark, S. (2021, October 7). *How Predictive and Prescriptive Analytics Improve the Call Center Experience.* CMSWire. https://www.cmswire.com/customer-experience/how-predictive-and-prescriptive-analytics-improve-the-call-center-experience/

70. Tyagi, A. (2023, December 12). *Top Contact Center Automation Trends for 2024.* Sprinklr. https://www.sprinklr.com/blog/contact-center-automation-trends/

Section 6. Talent and Expertise

71. Manning, H., & Bodine, K. (2012). *Outside In: The Power of Putting Customers at the Center of Your Business.* Houghton Mifflin Harcourt.

72. Abercrombie, C. (2023, August 22). *How to build and organize a machine learning team.* Enterprise AI. https://www.techtarget.com/searchenterpriseai/tip/How-to-build-and-organize-a-machine-learning-team

73. Brynjolfsson, E. (n.d.). *Applying AI: Building the organization for scaling AI*. Initiative for Applied Artificial Intelligence. https://aai.frb.io/assets/logos/AppliedAI_Whitepaper_OrganizingAI.pdf

Rule 30. The Role of Leadership

74. Infused Innovations, Inc. (2023, May 30). *Unleashing the Power of AI: Establishing an AI Center of Excellence*. LinkedIn. https://www.linkedin.com/pulse/unleashing-power-ai-establishing-center-excellence/

75. Vashishta, V. (2023, February 2). *What Is an AI Center of Excellence?*. AtScale. https://www.atscale.com/blog/what-is-an-ai-center-of-excellence/

76. Davenport, T. H., & Dasgupta, S. (2019, January 16). *How to Set Up an AI center of Excellence*. Harvard Business Review. https://hbr.org/2019/01/how-to-set-up-an-ai-center-of-excellence

Rule 31. Manage Your Project

77. *Identify the Different Stages of an Artificial Intelligence Project*. (2024, February 29). OpenClassrooms. https://openclassrooms.com/en/courses/7078811-destination-ai-introduction-to-artificial-intelligence/7161416-identify-the-different-stages-of-an-artificial-intelligence-project

78. Srivastava, S. (2023, August 24). *How to Manage AI Projects: From POV to Ready-to-Execute Solution*. Appinventiv. https://appinventiv.com/blog/ai-project-management/

79. Saltz, J. (2023, December 3). *7 Tips for AI Project Management*. Data Science Process Alliance. https://www.datascience-pm.com/ai-project-management/

Rule 32. Strategic Partnership

80. *TELUS International and Five9 Partner to Launch End-to-End Contact Center as a Service Featuring Powerful AI-Driven Insights*. (2023, November 8). Five9. https://www.five9.com/news/news-releases/telus-international-and-five9-partner-launch-end-end-contact-center-service

81. Gold, J. (2023, May 16). *Zoom announces partnership with Anthropic for AI call center services.* Computerworld. https://www.computerworld.com/article/3696278/zoom-announces-partnership-with-anthropic-for-ai-call-center-services.html

82. *Build vs. Partner vs. Buy: How to Choose the Right AI Strategy for Your Contact Center.* (n.d.). Level AI. https://thelevel.ai/resource/build-vs-partner-vs-buy-how-to-choose-the-right-ai-strategy-for-your-contact-center/

83. *What are the benefits of collaborating with vendors on innovation?.* (n.d.). LinkedIn. https://www.linkedin.com/advice/0/what-benefits-collaborating-vendors-innovation

84. Marr, B. (2023, November 6). *The Vital Role of Partnerships in Scaling Artificial Intelligence.* Forbes. https://www.forbes.com/sites/bernardmarr/2023/11/06/the-vital-role-of-partnerships-in-scaling-artificial-intelligence/?sh=314f8ffa7083

Rule 33. Contact Center Domain Knowledge

85. Lee, J. (2023, March 27). *How Can AI Fit into Customer Service Call Centers Effectively?.* Forbes. https://www.forbes.com/sites/forbes-businessdevelopmentcouncil/2023/03/27/how-can-ai-fit-into-customer-service-call-centers-effectively/?sh=5cbebb0f3fee

86. *Contact Center AI Bridges Gaps in Agent and Customer Connections.* (2023, June 7). Genesys. https://www.genesys.com/blog/post/contact-center-ai-bridges-gaps-in-agent-and-customer-connections

87. Mitchell, C. (2024, January 25). *Contact Center Generative AI: Use Cases, Risks, & Predictions.* CX Today. https://www.cxtoday.com/contact-centre/contact-center-generative-ai-use-cases-risks-predictions/

Rule 34. AI and ML Talent

88. *Understanding the business value of machine learning in AI.* (n.d.). Teradata. https://www.teradata.de/insights/ai-and-machine-learning/role-of-machine-learning-in-ai

89. Pratt, M. K. (2023, June 21). *12 key benefits of AI for business.* Enterprise AI. https://www.techtarget.com/searchenterpriseai/feature/6-key-benefits-of-AI-for-business

90. *How Do Businesses Use Artificial Intelligence?.* (2022, January 19). Wharton Online. https://online.wharton.upenn.edu/blog/how-do-businesses-use-artificial-intelligence/

Rule 35. Data Scientists and Analytical Skills

91. *How AI Is Rewriting the Rules of Data Analysis.* (2023, August 10). IIBA. https://www.iiba.org/business-analysis-blogs/how-ai-is-rewriting-the-rules-of-data-analysis/

92. Quantum Analytics Ng. (2023, September 5). *The Impact of Artificial Intelligence on Data Analytics.* LinkedIn. https://www.linkedin.com/pulse/impact-artificial-intelligence-data-analytics-quantum-analytics-ng/

93. *The Intersection of Data Analytics and Artificial Intelligence.* (2023, November 17). IABAC®. https://iabac.org/blog/the-intersection-of-data-analytics-and-artificial-intelligence

94. Chia, A. (2023, October). *5 Unique Ways to Use AI in Data Analytics.* Datacamp. https://www.datacamp.com/blog/unique-ways-to-use-ai-in-data-analytics

95. Arora, S. (2023, May 2). *AI analytics explained: How it works and key industry use cases.* ThoughtSpot. https://www.thoughtspot.com/data-trends/ai/ai-analytics

Section 7. Security and Privacy

Rule 36. Trust, Risk, and Security Management

96. Litan, A., D'Hoinne, J., & Willemsen, B. (2023, July 18). *Hype Cycle for the Future of Enterprise Applications, 2023.* Gartner. Retrieved January 13, 2024, from https://www.gartner.com/en/documents/4538599#:~:text=How%20does%20a%20Gartner%20Hype,or%20hanging%20on%20too%20long.

97. Perri, L. (2023, September 5). *Tackling Trust, Risk and Security in AI Models.* Gartner. https://www.gartner.com/en/articles/what-it-takes-to-make-ai-safe-and-effective

Rule 37. Threat Detection and Response

98. *AI Threats Detection.* (n.d.). DataDome. https://docs.datadome.co/docs/ai-detection

99. Grover, S. (2023, August 24). *Evolving from Threat Detection and Response to Threat Intelligence.* IDC. https://blogs.idc.com/2023/08/24/evolving-from-threat-detection-and-response-to-threat-intelligence/

100. *IBM Announces New AI-Powered Threat Detection and Response Services.* (2023, October 5). IBM. https://newsroom.ibm.com/2023-10-05-IBM-Announces-New-AI-Powered-Threat-Detection-and-Response-Services

101. Spring Rain Pvt Ltd. (2023, July 6). *The Role of Artificial Intelligence in Cybersecurity: Enhancing Threat Detection and Response.* LinkedIn. https://www.linkedin.com/pulse/role-artificial-intelligence-cybersecurity-enhancing-threat-1c

Rule 38. Manage Security Incidents

102. Padghan, V. (2024, January 25). *Role of Human Oversight in AI-Driven Incident Management and SRE.* Squadcast. https://www.squadcast.com/blog/role-of-human-oversight-in-ai-driven-incident-management-and-sre

103. Balroop, D. (2023, October 26). *The Role of AI in Data Privacy and Security in 2023.* LinkedIn. https://www.linkedin.com/pulse/role-ai-data-privacy-security-2023-dave-balroop-exite/

104. *Generative AI is shaping future incident management processes.* (2023, November 15). Help Net Security. https://www.helpnetsecurity.com/2023/11/15/incident-management-process-automation/

105. Ahmad, A. (2023, December 4). *From Detection to Resolution: AI in Incident Management.* Datafloq. https://datafloq.com/read/from-detection-to-resolution-ai-in-incident-management/

Rule 39. Vulnerability Management

106. Wang, B. (2023, April 17). *Vulnerability Management.* CrowdStrike. https://www.crowdstrike.com/cybersecurity-101/vulnerability-management/

107. Verma, A. K. (2024, January 10). *How AI is Transforming Vulnerability Management.* LinkedIn. https://www.linkedin.com/

pulse/how-ai-transforming-vulnerability-management-ashish-kumar-verma-n3kkc/

108. Robb, D. (2022, December 16). *Vulnerability Management as a Service (VMAAS): Ultimate Guide.* eSecurity Planet. https://www. esecurityplanet.com/networks/vulnerability-management-as-a-service/

109. R, R. K. (2023, June 29). *Vulnerability management using AI.* Beagle Security. https://beaglesecurity.com/blog/article/vulnerability-management-using-ai.html

Rule 40. Ethics and Customer Rights

110. Lawton, G., & Wigmore, I. (2023, October 10). *AI ethics (AI code of ethics).* TechTarget. https://www.techtarget.com/whatis/definition/AI-code-of-ethics

111. Wren, H. (2023, October 9). *Ethics of AI in CX.* Zendesk. https://www.zendesk.com/blog/ai-ethics-in-cx/

112. Kasai, R. (2023, December 7). *5 important considerations for responsible AI use in contact centers.* Talkdesk. https://www.talkdesk.com/blog/responsible-ai-use-in-contact-centers/

113. *Three Laws of Robotics.* (2024, April 9). Wikipedia. https://en.wikipedia.org/wiki/Three_Laws_of_Robotics

114. *Artificial Intelligence Act.* (2024, May 2). Wikipedia. https://en.wikipedia.org/wiki/Artificial_Intelligence_Act#cite_note-2

115. *Blueprint for an AI Bill of Rights.* (n.d.). The White House. https://www.whitehouse.gov/ostp/ai-bill-of-rights/

Rule 41. Privacy and Compliance

116. *Council of Europe Data Protection website.* (n.d.). Data Protection. https://www.coe.int/en/web/data-protection

117. Wilson, P. (2016, December 2). *Build Your Center-of-Excellence for Compliance Management.* LinkedIn. https://www.linkedin.com/pulse/build-your-center-of-excellence-compliance-management-phil-wilson-1/

118. Manuelap-Msft, Sericks007, KumarVivek, V-Amoldeore, Jenefer-Monroe, Mikefactorial, JeneferM-MSFT, DCtheGeek, Jimholtz, & Kathyos. (2023, September 6). *Use governance components.* Microsoft Build. https://learn.microsoft.com/en-us/power-platform/guidance/coe/governance-components

About the Author

Geoffrey A. Best started in the computer industry in the 1970s and has worked with contact centers for over 30 years. His career has taken him from computer-aided mapping and 911 emergency dispatch systems to computer telephony applications and today's contact center systems. Geoffrey has worked and consulted worldwide with utilities, communications, manufacturing, banking, and insurance companies. His experience has given him insight into companies' systems and methods to operate their contact centers and service their customers effectively. This experience has given Geoffrey a unique perspective on how customer expectations have changed over the past decades and how contact center solutions have evolved to satisfy them. His latest book introduces how the use of artificial intelligence will impact contact center operations and this new technology's impact on the customer experience.

www.ingramcontent.com/pod-product-compliance
Lightning Source LLC
Chambersburg PA
CBHW042117190326
41519CB00030B/7523